CONSTRUCTION SPECIFICATIONS PORTABLE HANDBOOK

Fred A. Stitt

McGRAW-HILL
New York San Francisco Washington, D.C. Auckland Bogotá
Caracas Lisbon London Madrid Mexico City Milan
Montreal New Delhi San Juan Singapore
Sydney Tokyo Toronto

Library of Congress Cataloging-in-Publication Data

Stitt, Fred A.
  Construction specifications portable handbook /
Fred A. Stitt.
      p.   cm.
  Includes index.
  ISBN 0-07-134103-X
  1. Buildings—Specifications—Handbooks, manuals, etc.
I. Title.
TH425.S73   1999
692'.3—dc21   99-14648
                                                      CIP

McGraw-Hill

A Division of The **McGraw·Hill** Companies

Copyright © 1999 by The McGraw-Hill Companies, Inc. Printed in the United States of America. Except as permitted under the United States Copyright Act of 1976, no part of this publication may be reproduced or distributed in any form or by any means, or stored in a data base or retrieval system, without the prior written permission of the publisher.

1 2 3 4 5 6 7 8 9   DOC/DOC   9 0 4 3 2 1 0 9

ISBN 0-07-134103-X

The sponsoring editor of this book was Wendy Lochner, and the production supervisor was Sherri Souffrance.

Printed and bound by R. R. Donnelley & Sons Company.

McGraw-Hill books are available at special quantity discounts to use as premiums and sales promotions, or for use in corporate training programs. For more information, please write to the Director of Special Sales, McGraw-Hill, 11 West 19th Street, New York, NY 10011. Or contact your local bookstore.

This book is printed on recycled, acid-free paper containing a minimum of 50% recycled de-inked fiber.

Information contained in this work has been obtained by The McGraw-Hill Companies, Inc. ("McGraw-Hill") from sources believed to be reliable. However, neither McGraw-Hill nor its authors guarantee the accuracy or completeness of any information published herein and neither McGraw-Hill nor its authors shall be responsible for any errors, omissions, or damages arising out of use of this information. This work is published with the understanding that McGraw-Hill and its authors are supplying information, but are not attempting to render engineering or other professional services. If such services are required, the assistance of an appropriate professional should be sought.

To the most unsung of unsung heros, the members of

the Construction Specifications Institute and

all the others who work so hard to

keep everyone in the industry

on their toes.

CONSTRUCTION SPECIFICATIONS PORTABLE HANDBOOK

CONTENTS

COMMON PROBLEMS WITH SPECIFICATIONS	1
HOW TO CREATE A SET OF SPECIFICATIONS	2
EFFICIENT WRITING	4
WHY WRITE SPECIFICATIONS AS CHECKLISTS?	5
HOW TO CREATE OFFICE MASTER SPECIFICATIONS	7
INSTRUCTIONS FOR USING THE CSI FORMAT	8
SPECIFICATION SECTION CHECKLIST SHORT-FORM	9
SPECIFICATION SECTION CHECKLIST LONG FORM	10
PROJECT MANAGER AND SPECIFICATIONS WRITER CONSIDERATIONS AND DECISIONS	12
SPECIFICATIONS ADMINISTRATION CHECKLIST	15
SPECIFICATIONS DEVELOPMENT	17
COMMON ERRORS & OMISSIONS IN SPECIFICATIONS	21
DON'T BE TAKEN IN BY BAD SUBSTITUTIONS DURING BIDDING	24

MASTERFORMAT DIVISIONS:

DIVISION 00	DOCUMENTS	25
DIVISION 1	GENERAL REQUIREMENTS	37
DIVISION 2	SITEWORK	53
DIVISION 3	CONCRETE	89
DIVISION 4	MASONRY	107
DIVISION 5	METALS	125
DIVISION 6	WOOD	148
DIVISION 7	THERMAL AND MOISTURE PROTECTION (WATERPROOFING & ROOFING)	167
DIVISION 8	DOORS AND WINDOWS	209
DIVISION 9	FINISHES	237
DIVISION 10	SPECIALTIES	296
DIVISION 11	EQUIPMENT	302
DIVISION 12	FURNISHINGS	309
DIVISION 13	SPECIAL CONSTRUCTION	315
DIVISION 14	CONVEYING	321
DIVISION 15	MECHANICAL	328
DIVISION 16	ELECTRICAL	344
APPENDIX -- SPECIFICATIONS DATA SOURCES		354

ABOUT THE AUTHOR

Fred A. Stitt is a registered architect and the founder and director of the San Francisco Institute of Architecture, an experimental graduate school focusing on organic, alternative, and ecological architecture. He also teaches extensively. He is the author of more than 50 manuals on architectural practice through his publishing company, Guidelines, and is the publisher and editor of *The Guidelines Letter*, a monthly newsletter for architects. Mr. Stitt has also written the *Working Drawing Manual*, the *Uniform Drawing Format Manual, Systems Drafting, Systems Graphics*, and the *Ecological Design Handbook*, all published by McGraw-Hill, in addition to several other quality control resources for architects and builders. A frequent speaker at AIA events and workshops on design practice, he gives lectures to thousands of architects in the United States and Canada each year. Mr. Stitt also taught at the School of Architecture at the University of California at Berkeley in the late 1980s and early 1990s.

COMMON PROBLEMS WITH SPECIFICATIONS

The most common problems with specifications are:

-- They too often include text from previous specifications that don't pertain to the project at hand.

-- They tend to be too wordy and difficult to read.

-- They sometimes contradict the working drawings. This especially occurs when working drawing notation is too elaborate. Sometimes notes include detailed material descriptions, product names, and quality standards, all of which are the province of specifications. In such cases, contradictions between drawing notation and specifications are virtually inevitable.

Here are a few rules most drafters, architects, and engineers know, but which sometimes drop out in actual practice.

1) Construction drawings show sizes, shapes, locations, and names of things, spaces, and materials. Drawings are the "first half" of construction documents and will frequently refer to specifications for elaborations on items named in the drawings.

2) Drawings provide identification notes that name generic descriptions of materials, such as "Concrete."

3) Specifications list products and detailed attributes of materials and may provide assembly instructions, trade and testing laboratory standards, and standards of installation and quality by which work will be judged.

4) Most specifications are not written as original documents. They're assembled from previous jobs, from master sets of specifications, and from trade association or manufacturers' specs.

5) Specifications may be proprietary, that is they will list specific products and manufacturers. Or they may be written to require a certain performance with final choice of products and manufacturers left to the contractor.

6) Specifications are legal documents, part of the total contract between the client ("Owner") and the contractor. The design professional acts as the agent of the Owner in assuring that the standards of the specifications are met by the contractor.

7) The professional society responsible for up-to-date standards, formats, and education in spec writing is the Construction Specifications Institute. You can't find a better source of standards and educational materials than the CSI, 601 Madison Street, Alexandria, VA 22314. 703-684-0300.

HOW TO CREATE A SET OF SPECIFICATIONS

HOW SPECS ARE "WRITTEN"

Specification writing is mainly specification assembly.

The architect or engineer doesn't create the most technical aspects of construction specifications. That information is created by experts on materials or by manufacturers of specific products.

The design professional is mainly a conduit who compiles the information, modifies it as necessary for the project, deletes the inapplicable data, and delivers it to the contractor for compliance.

This should reassure anyone who isn't well experienced in spec writing. You won't be expected to be an expert on everything, but you will be expected to learn how to find expert information and how to be sure that what you find is applicable to the job you're working on.

You may also have to restructure information to match your office's specifications format. That's mainly a matter of rearranging text rather than rewriting it.

THE FIRST STEPS

The first step is to assemble all project-related data. Often this isn't done until working drawings are near completion, but it's excellent policy to establish a project specifications binder early in the working drawing phase.

The binder will contain references to all specifiable items and may include mini-sets of progress drawings, to help identify everything that's going into the building.

With this information, the spec writer can create a master list of specifiable items. This will be a Table of Contents that will serve as a checklist of what's to be included in the job.

If the project manager is not writing the specifications, he or she should notify the specifications writer of all tentative and final decisions on materials and products to be used in the project. That allows the spec wrier some time to gather up-to-date manufacturer and trade association literature.

As noted above, someone new to specifications writing needs to know that most of the data he or she will use is already written. The overall outline will most likely follow that of the Construction Specifications Institute Masterformat. Concrete standards will be as provided by the American Concrete Institute. And so on.

Much of the data may also be provided by a commercial master specification such as MasterSpec by the CSI and Quick Specs by Guidelines, etc.

In many offices, the bulk of data may well come from previous similar projects or from a customized office set of master specifications.

So, as you see, there's plenty of ready made data to work with. Wherein lay pitfalls and danger spots. The data all has to be checked and cross checked for relevance to the project, for congruence with the working drawings, and for accuracy and timeliness of the trade reference standards. This last part is the one least carefully checked in design offices.

Many trade standards for material manufacture are listed in the ASTM manuals, but most firms don't have copies of those manuals, or the ones they have are out of date.

Your best source of up-to-date data will be from CSI data sheets, professionally written specifications sections on numerous aspects of construction.

THE FIRST DRAFT

For the first draft of specifications, the writer will create computer files of all the major divisions and subdivisions of a standard set of specs. These may include many subdivisions that will NOT be in the project.

The basic structure of information will follow the 16 divisions of the CSI Masterformat. (If you make a project binder, you'll have index tabs which match these divisions: 00 for Documents, 01 for General Requirements, 02 for Sitework, 03 for Concrete, and so forth.)

If the master specifications you're working with include data on concrete slabs and slabs are part of the project, then you'll keep that data. If the master specs include data on asphalt paving that are not in the project, you'll delete that data.

So, in fairly short order, most of the existing data will be assembled, and if the sources are up to date, the trade standards and references will be OK.

WRITING NEW SPEC SECTIONS

What do you do if you have to write a new section of specifications from scratch?

Just follow the formats and outlines shown on the pages that follow. Then follow the writing style instructions, and start each sentence with phrases like: "The Contractor shall ..." and then use other similar texts as a guide. It's like putting together a puzzle that has pieces missing. So first you find what's missing in the information that you'll communicate on the specs, then find out where to get it.

EFFICIENT WRITING

Examples of superfluous phrases to watch for when writing and editing include:

"The contractor shall furnish and install..." which could be written: *"Provide..."**

The phrases:

"As manufactured by," "as indicated," "in accordance with"

could be written:

"Per."

Filler words that can often be eliminated with no loss in meaning include:

"The," "of," "and," "a," "all," "every," "which," "shall be," "will be."

Generalities and subjective terms should be struck out or replaced with fewer, more specific words. Such terms add excess wordage, and force bidders to add contingency costs to cover the unpredictability of future interpretations. Examples are:

"In the opinion of," "to the satisfaction of," "in an approved manner," "as required," "in accordance with good practice," "in a neat and workmanlike manner," "highest quality," "first class."

Hazardous words that may permit varying interpretation within a particular context include:

"Any," "other," "etc.," "and/or."

What follows are some sentences taken from fairly average specs. Each is rewritten to illustrate one or more of the many time and space savers available to spec writers.

Original sentence:

"The contractor shall verify all dimensions and conditions in the field and shall prepare detailed shop drawings of duct work for the Architect's approval."

Rewritten as two sentences, this becomes:

"Verify all field dimensions and conditions. Provide detailed duct work shop drawings for Architect's approval."

---

*It is imperative to define within a legend in working drawings or specs the use of "provide" or any other short term, symbol or abbreviation that encompasses longer terms or phrases.

Original sentence:

"Sliding Aluminum Doors. Contractor shall furnish and install sliding aluminum doors, 'Elmo,' Series 50, as manufactured by E. A. Co., Flint, Michigan."

Written in the recommended "streamlined" form, this would be:

"Sliding Aluminum Doors. 'Elmo,' Series 50, E.A. Co., Flint, Mich."

Wordage is cut by one-half.

Frequently a dozen words can be condensed into two words, as in this example under "Framing Lumber":

"Allowable moisture content. All framing lumber shall have a moisture content of not greater than 20%."

could have been written:

"Allowable moisture content: 20% max."

Some of the problems of referencing to building codes, trades, and testing lab standards are cited elsewhere. An example of a missed opportunity for using references is seen in the following:

"Bridging. All stud partitions or walls over ten feet (10') in height shall have bridging, not less than two inches (2") in thickness and of the same width as the stud, fitted snugly and spiked into the studs at their midheight, or other means for giving lateral support to the studs, or as per local ordinance."

This could have been:

"Bridging: per local ordinance."

or:

"Bridging: UBC, 2507 (b), 1998 ed."

And a final note, contributed by an English major-turned professional spec writer: *"Write simply. This means avoid all compound sentences. Include only one indivisible idea per sentence. And include only one indivisible topic per paragraph. The greater the number of short sentences, the shorter the spec. The shorter the spec, the more likely it is to be read, understood, and used."*

WHY WRITE SPECIFICATIONS AS CHECKLISTS?

When you make a shopping list, do you lump everything together in a single paragraph?

In writing a To-Do list of calls to make and appointments to keep, do you string items together in long sentences?

Not likely.

So why write specifications that way?

Specifications are shopping lists. And To-Do lists.

For years specifications have been written in paragraph form in a style that blends the worst aspects of legal contracts and technical manuals. Despite years of effort at simplification, such as Outline Specifications, Streamlined Specs, and Short Form Specs, most A/E's still write them as if they were law books.

Here's a sample of a specification section from a published master specification:

"Provide temporary piping and water supply, and, upon completion of the work, remove such temporary facilities."

Rewritten in checklist format, it looks like this:

__ Provide temporary:
    __ Piping
    __ Water supply

__ Remove these upon completion of work.

Here's another example of standard spec writing:

"Provide and maintain for the duration of construction all scaffolds, tarpaulins, canopies, warning signs, steps, platforms, bridges, and other temporary construction necessary for the completion of the work in compliance with pertinent safety and other regulations."

First we mark the "checklist" items with asterisks, like this:

"Provide and maintain for the duration of construction all *scaffolds, *tarpaulins, *canopies, *warning signs, *steps, *platforms, *bridges, and *other temporary construction necessary for the ... etc."

Then we rewrite it in the checklist format:

__ Provide and maintain for duration of construction:

    __ Scaffolds
    __ Tarpaulins
    __ Canopies
    __ Warning signs
    __ Steps
    __ Platforms
    __ Bridges
    __ Any other temporary construction necessary for completion of the work

__ Temporary construction shall comply with:

    __ All safety regulations
    __ All other applicable codes

The checklist version is as complete as the first version, but much easier to read and understand.

Best of all, the contractor (and you) can check off the listed items, one by one, to be sure they're actually done on the job as you require.

HOW TO CREATE OFFICE MASTER SPECIFICATIONS

Regardless of what published master specifications or specs from previous project you use as resources, every office must create its own generic master specifications. Here are the steps:

__ If you haven't already done so, acquire CSI literature on the 16 Division Masterformat system. This literature can be obtained from the Construction Specifications Institute (CSI, 601 Madison Street, Alexandria, VA 22314. 703-684-0300).

__ Acquire pertinent reference sources and master specification texts listed in the Appendix, as you need them.

__ Rewrite existing specifications text in checklist format.

__ When writing your next set of specifications, use the topic checklists in this handbook as reminders of items you may want to add.

__ Add your own ideas of reminders to this checklist for future reference.

__ Create file folders and/or three-ring binders to compile copies of your new standard texts.

Notes:

INSTRUCTIONS FOR USING THE CSI FORMAT

Here are rules and definitions from the Construction Specifications Institute Masterformat. They are a great help in writing complete, well-coordinated specifications.

"Divisions" are the 16 main categories of the CSI Masterformat system of organizing specifications.

"Sections" include specific requirements for units of work within a Division.

"Parts" are three main subdivisions of sections as listed below.

"Articles" consist of related paragraphs within a Part of a Section.

"Paragraphs" can be single sentences or groups of sentences dealing with a single product, standard, measurement, etc.

The **"Parts"** of a Section are as follows:

PART 1--GENERAL

> Deals with administrative data and procedures, references, submittals, quality control, project conditions, warranties, and maintenance.

PART 2 -- PRODUCTS AND MATERIALS

> Provides explicit directions as to types, names, manufacture and fabrication.

PART 3--EXECUTION (or INSTALLATION)

> Site and work preparation, installation or applications, relationships with other products or materials, standards for workmanship, standards for inspection, final corrections,

All three parts are described in short-form and long-form versions on the next few pages.

Notes:

_____

_____

_____

SPECIFICATION SECTION CHECKLIST SHORT-FORM

PART 1 -- GENERAL

___ Administrative data and procedures
___ Summary
___ References
___ Submittals
___ Quality Assurance
___ Delivery, Storage, and Handling
___ Project/Site Conditions
___ Sequencing and Scheduling
___ Warranty
___ Maintenance

PART 2 -- PRODUCTS AND MATERIALS

___ Types
___ Names
___ Manufacturers
___ Component(s)
___ Accessories
___ Mixes
___ Fabrication
___ Manufacturer or Supplier Quality Control

PART 3 -- EXECUTION (or INSTALLATION)

___ Examination/Verification of Conditions
___ Preparation/Protection
___ Erection/Installation/Application
___ Interface with Other Products
___ Workmanship
___ Standards

Notes:

_____

_____

_____

_____

_____

_____

_____

_____

SPECIFICATION SECTION CHECKLIST   LONG FORM

Use this CSI list as a reminder checklist on general completeness of your specification sections. This follows the almost universally accepted "Three-Part" format.

PART 1--GENERAL

    ___ Summary
    ___ Administrative Data and Procedures
    ___ Section Includes
    ___ Products Furnished But Not Installed
    ___ Products Installed But Not Furnished Under This Section
    ___ Related Sections
    ___ Allowances
    ___ Unit Prices
    ___ Alternates/Alternatives
    ___ References
    ___ List of Standards (By Acronym and Alphanumeric Identification such as ASTM)
    ___ Definitions

    ___ System Description
        ___ Design Requirements
        ___ Performance Requirements

    ___ Submittals
        ___ Product Data
        ___ Shop Drawings
        ___ Samples

    ___ Quality Control Submittals
        ___ Design Data
        ___ Test Reports
        ___ Certificates
        ___ Manufacturers' Instructions
        ___ Manufacturers' Field Reports

    ___ Contract Close-out Submittals
        ___ Project Record Documents
        ___ Operation and Maintenance Data
        ___ Warranty

    ___ Quality Assurance
        ___ Qualifications for Contractor-Employed Designers
        ___ Regulatory Requirements
        ___ Certifications
        ___ Field Samples
        ___ Mock-Ups
        ___ Pre-Installation Conference

    ___ Delivery, Storage, and Handling
        ___ Packing and Shipping
        ___ Acceptance at Site
        ___ Storage and Protection

    ___ Project/Site Conditions
        ___ Environmental Requirements
        ___ Existing Conditions
        ___ Field Measurements

__ Sequencing and Scheduling

__ Warranty
    __ Special Warranty

__ Maintenance
    __ Maintenance Service
    __ Extra Materials
    __ Items furnished to the owner by contractor for future maintenance and repair

PART 2--PRODUCTS AND MATERIALS

__ Manufacturers/suppliers
__ Materials
__ Manufactured Units
__ Equipment
__ Component(s)
__ Accessories
__ Mixes
__ Fabrication
    __ Shop Assembly
    __ Shop/Factory Finishing
    __ Tolerances

__ Supplier Quality Control
    __ Tests
    __ Inspection
    __ Verification of Performance

PART 3--EXECUTION (or INSTALLATION)

__ Examination
    __ Verification of Conditions

__ Preparation
    __ Protection
    __ Surface Preparation

__ Erection/Installation/Application
    __ Manufacturers' Instructions
    __ Special Techniques
    __ Interface with Other Products
    __ Tolerances

__ Field Quality Control
    __ Tests
    __ Inspection
    __ Manufacturers' Field Service

__ Adjusting
__ Cleaning
__ Demonstration
__ Protection
__ Schedules

PROJECT MANAGER AND/OR SPECIFICATIONS WRITER CONSIDERATIONS AND DECISIONS

A set of specifications, like every other aspect of design services, requires a plan of implementation. Implementation is mainly a matter of making choices.

The checklist that starts below includes the most common and crucial decisions regarding the creation of any set of specifications.

The amount of work involved depends on the size, budget, and required attention to detail, which allows services to be divided as:

1) "Baseline" -- the essential services required for a construction permit and owner-built or negotiated contract work.

2) "Standard" -- the range of services normally included in the AIA contract for services.

3) "Extended" or fully comprehensive services that include more than the normal range of design service.

The specification writer needs to be fully aware of the design services contract, construction budget and category (economy, mid-level, and highest quality), and design service fees, in order to judge how detailed and comprehensive the specifications should be.

In that regard, the basic issues cited in these lists should be part of the spec writer's awareness and planning to-do list.

BASELINE SERVICES

___ If the client is also the contractor, no specification may be required

___ For small projects, specifications may be handled as general notes in the drawings

___ For larger jobs, the design firm may have to review the following decisions and options with the client:
  ___ Decide with client the type of construction contract
  ___ Decide with client the contractor fee types

___ Confirm the type of specification

Notes:

_____

_____

_____

STANDARD SERVICES TASKS AND DECISIONS

Includes tasks required for larger projects as well as small ones. The list should be edited accordingly.

__ Conduct review meetings and/or drawing checks to coordinate decisions and alternatives on:
  __ Room functions and relationships
  __ Room finish schedule
  __ Construction system
  __ Structural system
  __ Mechanical system
  __ Lighting
  __ Vertical transportation

  __ Exterior materials:
    __ Roofing
    __ Walls
    __ Fenestration

  __ Interior partitioning system
  __ Cabinetry
  __ Site appurtenances

  __ Materials, finishes, and fixture quality:
    __ Superior
    __ Middle Grade
    __ Economy Grade
    __ Mixed Grades

__ Identify and write specification sections that can be completed early in the working drawing process

__ Make checklist of latest special trade association standards required for the project

__ Confirm use of latest applicable product literature

__ Confirm use of latest applicable testing agency standards

__ Confirm use of all latest applicable codes and regulations

__ Acquire from the client any previous relevant specifications

__ Create a Project Manual binder for preliminary organization of specification information (use index tabs following the CSI Masterformat)

EXTENDED SERVICES TASKS AND DECISIONS

Includes tasks required for larger projects as well as small ones.

__ Provide completely detailed specification divisions and sections incorporating the following data:

  __ Materials:
    __ Generic name
    __ Proprietary name with manufacturer
    __ Description by use
    __ Description by performance criteria
    __ Description by reference standard

- ___ Required characteristics of materials:
    - ___ Gauge or weight
    - ___ Sizes, nominal or finished
    - ___ Type of finish
    - ___ Allowable moisture content

- ___ Components or proportions of components of materials:
    - ___ Mixes
    - ___ Care in handling
    - ___ Temperature protection
    - ___ Moisture protection

- ___ Installed location on the job if not fully indicated in the drawings

- ___ Preparation for installation:
    - ___ Pre-job inspection
    - ___ Coordination with other subcontractor(s)
    - ___ Cleaning
    - ___ Preparation of surfaces

- ___ Installation:
    - ___ On-site fabrication
    - ___ Connection to other work
    - ___ Adjusting and fitting
    - ___ Finishing

- ___ Coordination:
    - ___ Broadscope working drawing sheet reference
    - ___ Detail drawing sheet reference
    - ___ Consultant's drawing sheet reference
    - ___ Related and/or connecting work by other trades or subcontractors
    - ___ Related other specifications sections

- ___ Workmanship standards and tolerances:
    - ___ Quantified measurements
    - ___ Referenced to published standards
    - ___ Approval by inspection

- ___ Inspections and tests (may be combined with workmanship standards and tolerances)

- ___ Repair and patching

- ___ Clean-up, preparation for other work

- ___ Warranties, bonds, or guarantee requirements

- ___ Postconstruction adjustments or service

___ Review "Scope of Work" and "Work Not Included" articles in each section

___ Verify all references to work in other sections

___ Review the Special Conditions

___ Distribute copies of specifications for content review by department heads and/or job captains, and the designated project site representative(s)

SPECIFICATIONS ADMINISTRATION CHECKLIST

DECISIONS TO DOCUMENT

All design services contract decisions must be explicit and fully understood by the specifications writer.

These include:

__ Confirm the type of construction contract:

- __ Competitive Bidding -- Open
- __ Competitive Bidding -- Selected Contractors.
- __ Negotiated Contract.
- __ Single Prime Contract.
- __ Multiple Separate Contracts.
- __ Stipulated Lump Sum. (Refer to AIA Handbook)
- __ Cost Plus Fee. (Refer to AIA Handbook)
- __ (other)

__ Fee types:

- __ Fixed Fee.
- __ Fixed Fee with Guaranteed Maximum.
- __ Percentage of Construction.
- __ (other)

__ Related options:

- __ Phased Construction.
- __ Fast Track.
- __ Construction Management.
- __ Design-Build.
- __ Contractor prepared construction documents
- __ (other)

__ Confirm the type of specification. (Refer to AIA Handbook)

- __ Open/Contractor's Option.
- __ Closed/Proprietary.
- __ Product Approval.
- __ Substitute Bid.
- __ Approved Equal.
- __ Product Description.
- __ Performance.
- __ Work Procedure.
- __ (other)

__ Decide the specifications page numbering system. (Recommended: Number by division and division page number. Show start and finish page numbers for each division in the Table of Contents or Index.)

__ Establish type style, headings, and line indentation standards. (Recommended: Identify the project and design firm at the top of each page.)

- Decide on any special reprographic combination of specifications with working drawing sheets:
    - Specifications printed on working drawing size sheets and included as the final sheets.
    - Specifications bound with matching working drawing divisions.
    - Specifications bound with a detail book.
    - (other)

- Review the design and production schedule to identify the best times for drawing/specification coordination checks and meetings.

- Printing decisions:
    - Print on one or both sides of each sheet.
    - Paper -- type, quality, color code for major divisions.
    - Duplication method.
    - Binding -- fixed, loose-leaf, loose-leaf with sections fastened together.
    - Quantity.
    - (other)

Notes:

SPECIFICATIONS DEVELOPMENT

PREDESIGN AND SCHEMATIC DESIGN PHASES

All activities involved in specifications writing and coordination should occur in a logical start-to-finish sequence and be fully coordinated with all participants. The checklist below lists the main steps within each major phases of design work

__ Schedule review meetings and/or drawing checks to coordinate decisions and alternatives on:

   __ Room functions and relationships.
   __ Construction system.
   __ Structural system.
   __ Mechanical system.
   __ Lighting.
   __ Dominant exterior materials.
   __ Interior partitioning system.
   __ (other)

   __ Overall materials, finishes, and fixture quality.

      __ Superior.
      __ Middle Grade.
      __ Economy Grade.
      __ Mixed Grades.

DESIGN DEVELOPMENT PHASE

__ Schedule review meetings and/or drawing checks to coordinate decisions on:

   __ Preliminary room finish schedule.
   __ Construction system.
   __ Structural system.
   __ Mechanical system.
   __ Lighting.
   __ Vertical transportation.
   __ Exterior materials.

      __ Site appurtenances.
      __ Roofing.
      __ Walls.
      __ Fenestration.

   __ Interior partition systems.
   __ Cabinetry.
   __ Specific area materials, finishes, and fixture quality.

      __ Superior.
      __ Middle Grade.
      __ Economy Grade.
      __ Mixed Grades.

RESEARCH AND DATA GATHERING

__ Schedule review meetings and/or drawing checks to coordinate decisions on all construction materials and systems. (Recommended: Use the design development room finish schedule as a preliminary guide.)

__ Identify specification sections that can be completed early in the working drawing process.

__ Start a checklist of special standards and product literature required for this project.

__ Call or send for latest applicable product literature. See Sweet's Selection data file.

__ Send for the latest applicable testing agency standards.

__ Acquire copies of all applicable codes and regulations.

__ Acquire copies of previous relevant office specifications.

__ Contact the client for previous relevant specifications.

__ Create a Project Manual binder for preliminary organization of specification information. Use index tabs following the CSI Masterformat.

__ Create a master list of items to be specified. A list can be copied from the CSI Masterformat subdivisions.

__ If using a master guide specification such as Masterspec, Spectext, Guidelines QuickSpecs, etc., match their completed sections to your master list. Note items to be covered that are not in the master guide specifications.

__ Identify specification sections that can be written or assembled first without extensive working drawing development.

__ Acquire printout or photocopy of overall master guide specification sections.

__ Cut and paste (physically or on computer) the preliminary reference or master guide specifications and add further notes as the guide for the typist or computer operator.

FINISH SPECIFICATIONS:
WRITING AND CHECKING CHECKLIST

__ Verify that specification section numbers consistently follow one system or one edition of a system such as the CSI Masterformat.

__ Assign proofreading and establish a proofreading system section by section. (Recommended: One person reads the last original draft aloud while another person reads and compares the verbal reading with the finished printout. This is the best way to catch errors and data drop outs.)

__ Check the completeness of specification sections. Check the consistency of the sequence of information in different sections. Items options, and sequence to check:

   __ Materials.
      __ Generic name.
      __ Proprietary name with manufacturer.
      __ Description by use.
      __ Description by performance criteria.
      __ Description by reference standard.

   __ Required characteristics of materials.
      __ Gauge or weight.
      __ Sizes, nominal or finished.
      __ Type of finish.
      __ Allowable moisture content.

   __ Components or proportions of components of materials.
      __ Mixes.
      __ Temperature protection.
      __ Moisture protection.

   __ Installed location on the job if not fully indicated in the drawings.

   __ Preparation for installation.
      __ Pre-job inspection.
      __ Coordination with other subcontractor(s).
      __ Cleaning.
      __ Preparation of surfaces.

   __ Installation.
      __ On-site fabrication.
      __ Connection to other work.
      __ Adjusting and fitting.
      __ Finishing.

   __ Coordination.
      __ Broadscope working drawing sheet reference.
      __ Detail drawing sheet reference.
      __ Consultant's drawing sheet reference.
      __ Related and/or connecting work by other trades or subcontractors.
      __ Related other specifications sections.

- Workmanship standards and tolerances.
    - Quantified measurements.
    - Referenced to published standards.
    - Approval by inspection.

- Inspections and tests. (May be combined with workmanship standards and tolerances.)
- Repair and patching.
- Clean-up, preparation for other work.
- Warranties, bonds, or guarantee requirements.
- Postconstruction adjustments or service.

___ Read "Scope of Work" and "Work Not Included" articles in each section.

___ Verify all references to work in other sections.

___ Review the Special Conditions.

___ Distribute copies of specifications for content review by department heads and/or job captains, and the designated project site representative(s).

___ Check all final sets of printed specifications for incorrect collating, missing pages, and printing flaws, or blanks.

Notes:

COMMON ERRORS & OMISSIONS IN SPECIFICATIONS

The spec writer should be fully aware of certain types of common errors, omissions, and lesser but notable quality control issues.

Contractors describe the following items as the sources of the most common questions and trouble spots they find in specifications. The list should be kept handy for review throughout the spec writing process, and the spec writer should add his or her own list.

__ What is the precise procedure for getting approval of substitutions?

__ When separate contracts are involved, who pays for:
   __ Heaters.
   __ Emergency weather protection.
   __ Storage sheds and platforms for subcontractors.
   __ Cleanup and removal of subcontractors' trash.
   __ Removal of faulty equipment.

__ Who sets survey marking lines, base lines, and elevations?

__ Who provides locations of utility hook-up points?

__ Does a cash allowance cover purchase cost only, or does it include related costs of delivery, unloading, and storage?

__ When you say "work by others," which "others" do you mean?

__ When trades and subcontractors' work overlap, such as when tile work is specified as part of carpentry, or painting of equipment is part of HVAC, which subcontractors are responsible and which are not?

__ When equipment supports and anchors are specified for walls, floors, and ceilings, which subcontractor is responsible?

__ Who pays for tests and special inspections?

__ If tests are specified to be conducted in the presence of design firm representatives and they don't show up as scheduled, can that be taken as approval to proceed with testing without the observers?

__ When standards publications are identified as "the latest edition," does that mean the latest edition when the specifications were written, or the latest during bidding?

__ If a reference standard contradicts an article in the General Conditions, which should govern?

__ Is there a cut-off date when no further addenda will be sent out?

__ Is there a single list of all cash allowances that are scattered throughout the specifications?

__ Is there a single list of N.I.C. items, and items provided by Owner but installed by Contractor?

__ When specifying a new or unusual product or material, what is the name, address, and phone number of the source?

And here are some general suggestions from builders and work crews:

It is very confusing when an architect incorporates the standard AIA "General Conditions" and then amends most of it. If there are that many amendments, it would be easier just to rewrite it.

When scheduling a bid opening, watch out for other local big projects that might be due at the same time. The existence of other work, and the extra rush of work in bidding, will push bids higher. (Tuesday through Thursday afternoons are the best times for contractors to get their bids in.)

Last-minute addenda are useless and confusing; bids go up when these come in. Set a cut-off date when there will be no further addenda, preferably within a week of the bid opening.

Don't bother sending addenda that only note slight dimension changes and minor corrections having no effect on costs.

If tight bids are desired, allow generous time for the bid period. (Three to four weeks is suggested as fair for medium to large jobs, six weeks for a large hospital.)

A completion date is meaningless without specified damages for non-compliance. Specified damages are meaningless without a reward for early completion.

Workmanship tolerances should be realistic; some are set with no idea of the limits of normal construction practice.

It helps to include a single, complete list of all the cash allowances scattered throughout the sections, all the work N.I.C., and a Drawing Index.

A Subject Index after the Table of Contents helps the subs find little odd jobs that appear in unlikely places. You can use your own spec checklist or the CSI outline, and add the appropriate page numbers after each item.

When using allowances, it helps to give enough design information so labor required for installation can be estimated.

Use of a hardware allowance makes bidding easier, and gives the architect extra time to make decisions after the job is out.

It is good to divide the spec sections by trades, but some architects don't bother to check out jurisdictional differences in the area they're designing for. For instance, in some cities the plumber installs everything in a bathroom, including the robe hook.

Don't bother asking the contractor to point out discrepancies under threat of penalty for failure to do so -- it isn't enforceable.

When excessive penalty clauses are included, some contractors just add estimated penalties onto the bid; the owner ends up paying for them.

Everyone gets a better deal with the base bid system, combined with a means of offering substitution proposals that spell out savings to the owner. It doesn't help to say that substitutions will be considered if the item is really closed. If worried about accusations of collusion with a manufacturer, build a list of favored comparable products and rotate their names from job to job.

The electrical and mechanical specs are usually the worst on the job. Why not check, and coordinate them with the architectural specs, and get rid of all the "General" and "Special Conditions" that either duplicate or conflict with the architectural specs?

If mentioning some new product, it helps a lot to include the name, address, and phone number of the local representative.

Notes:

DON'T BE TAKEN IN BY BAD SUBSTITUTIONS DURING BIDDING

This is a well-known troublespot, but one of the best solutions isn't widely practiced. First the problem:

Contractors can often improve on your drawings and specifications. They may have access to non-specified products that are cheaper and as good as, or better, than items you're familiar with. So it's reasonable to use the "or equal" phrase in specs, to allow for that likelihood.

On the other hand, "or equal" opens you up to chiseling. Low-ball bidders specialize in coming up with low-quality substitutions. It's during the latter phases of bidding, and during construction that ill-informed, flawed, last-minute decisions are likeliest to be made by A/E staff.

The "or equal" risk factor goes way up if you, your client, or any of your staff accept bidder substitutions that the other bidders don't know about. At that point, you are properly subject to lawsuit by contractors who didn't have a chance to offer similar substitutions and bid on the same basis.

Some design firms enforce a procedure that cuts through the dilemma nicely. The key is a special section on "approvals" that is included in the General Conditions of the contract. It says the following:

1) The specifications name the materials, products, and manufacturers required for the job.

2) Bidders may submit other materials, products, etc. for consideration as substitutions.

3) Submittals of proposed substitutions must be:

    -- In writing.
    -- Received ten (or other specified number of) days before the bid opening.
    -- Made in good faith, that is, verifiably equal or superior to the specified items.

4) Submittals must include all the data that would be in construction drawings and specifications:

    -- Complete names and descriptions.
    -- Dimensions.
    -- Performance figures.
    -- Latest catalogue numbers.

5) If new materials are named, data are provided on required laboratory tests, standards, etc.

6) If a new fabricator is named, data are required on capabilities and experience.

7) If the design firm approves a bidder substitution submittal, copies of the submittal will be delivered to all other bidders so that all bids reflect the same options.

DIVISION 00
DOCUMENTS

DOCUMENTS
00000

CONTENTS

00010	INVITATION TO BID	26
00100	INSTRUCTIONS TO BIDDERS	27
00300	EXAMPLE OF A BID FORM	29
00700	GENERAL CONDITIONS OF THE CONSTRUCTION AGREEMENT	32

*Note to Specifier: Agreement Forms, Bond Forms, etc. which are normally a part of Division 00, should be as available from the AIA or from a qualified attorney. Other legal documents included in Divisions 00 and 01 of these checklists are for reference purposes only. Specification documents are usually preceded with a Title Page and Table of Contents.

00010 -- INVITATION TO BID

1. The Owners (*Note to specifier: Add name of the Owner(s), and/or Owner company or institution name.)

will accept bids for the construction of (*Note to specifier: Add a descriptive name and address of the project.)

2. The bid will be for all work as shown in Drawings and Specifications and related work required for project completion.

3. All bidders must examine the Drawings, read the Specifications, and visit the site of this project to fully investigate the extent and quality of the work required. Bidders shall be familiar with the location and access to construction site, availability of utilities, the condition of the site and any existing construction, and governing regulatory agencies and permit processes.

4. The successful bidder shall furnish the Owner with certificates of insurance in amounts listed below or other amounts as required by law, whichever is greater:

*Note to specifier: Confirm the amounts of insurance appropriate to the size and type of the project.

Workmen's Compensation Insurance for at least $500,000 each occurrence and $500,000 total for Bodily Injury including Personal Injury.

Property Damage for at least $100,000.

Comprehensive Automobile Liability for at least $250,000 for each person, $500,000 each occurrence, and Property Damage for at least $100,000 for each accident.

5. Submit bids on the forms provided, signed, with all items complete.

6. Address bids to the Owners and deliver to the address on the INSTRUCTIONS TO BIDDERS on or before the day stated.

_____
Owner(s) name and title

_____

(date)

END OF DOCUMENT

00100 -- INSTRUCTIONS TO BIDDERS

1. Project name and Owner's name:

2. This document contains Instructions to Bidders for the project named above.

3. To obtain bid documents, contact:
(*Note to specifier: Include name, firm name, street address, city, state, zip, phone, fax, and e-mail.)

4. A deposit of $ is required to obtain a copy of bidding documents. The deposit will be returned when the documents are returned complete and reusable within 7 calendar days after bid due date.

5. Submit Document 00300 -- Bid Form by: (*Note to specifier: State day of the week, date, and time of day.)

Submit bids in sealed envelopes labeled with project name and bidder's name. Mark the envelope: "Bid Enclosed. Do Not Open before Bid Time and Date." Late submissions will not be considered.

Submit bid to:
(*Note to specifier: Add firm name, street address, city, state, zip, and phone.)

6. Bid Security is required equal to (%) of the bid.

Bid Security must be as per AIA bid bond, paid as a certified or cashier's check made payable to the Owner.

Bid security will be forfeited if a bidder is awarded the contract and fails to complete the Owner-Contractor Agreement within 10 days of notification by Owner.

Bid Security for unsuccessful bidders will be returned within 14 calendar days after the contract is awarded.

7. Each bidder shall submit proof of bonding for the entire cost of the work. Bond form shall be as per AIA bid bond. Bonds must be by a surety company acceptable to the Owner and licensed in the locale of the work.

8. No modifications to bids will be considered except as included in the sealed bid envelope.

9. The Owner reserves the right to select or reject bids on the basis of price considered along with contractor's experience, location, reputation, status with contractor licensing regulators, and other qualifications. The Owner reserves the right to enter negotiations with any bidder after the bid opening date.

10. The Owner reserves the right to modify the Contract Documents and restart the bid process as necessary to meet the construction budget.

11. The bidder must review all documents with care to fully understand all the conditions of the project.

12. Questions may be submitted during the bidding period. Questions will be answered in writing in a timely manner and copies will be distributed simultaneously to all bidders. Submit questions to: (*Note to Specifier: Include name, firm name, street address, city, state, zip, phone, Fax, and
e-mail.)

13. A site visit for review of existing conditions is required. The bidder will be assumed to have full knowledge of site conditions that might affect the bid and subsequent construction costs. Contact the person named above to arrange to visit the site.

END OF DOCUMENT

Notes:

00300 -- EXAMPLE OF A BID FORM

*Note to specifier: Provide a Bid Form, preferably using AIA forms depending on the contract type such as Multiple Separate Contracts, Stipulated Lump Sum, Cost Plus Fee or any other variation required for the project. This example for Stipulated Sum is for reference only.

To (Owner):

1. The undersigned, has examined the construction documents including Drawings and Specifications, visited the site and examined the conditions affecting the work, hereby proposes and agrees to furnish all labor, materials, equipment and to perform all work necessary to complete the project as required by the construction documents, for the stipulated sum of:

_____ Dollars ($ _____).

2. The undersigned agrees that, if selected to perform the work, will, within (5) working days after notification by the Owner, sign a contract in accordance with the terms of this bid and comply with the Instructions to Bidders.

3. The undersigned agrees to maintain and hold the contract price stated in this proposal for a period of thirty (30) days from the date of this proposal.

This agreement entered into this ___ day of _____ , (year) _____ .

Submitted by:

END OF DOCUMENT

Notes:

_____

_____

_____

EXAMPLE OF A BID FORM

*Note to specifier: Provide a Bid Form, preferably using AIA forms depending on the contract type such as Multiple Separate Contracts, Stipulated Lump Sum, Cost Plus Fee or any other variation required for the project. This example is for reference only.

1. Submit bids in compliance with Document 00100 - Instructions to Bidders. The Owner reserves the right to reject Incomplete bid forms.

Name and Address of Bidder:

2. Base Bid.

The Bidder will perform all Work required by the Contract Documents for the amount of:

(Numerical bid) $

(Written bid)   dollars

3. Alternates.

If an Alternate is selected by the Owner, the Bidder proposes to do the Work required by the Contract Documents by increasing or decreasing the Base Bid the following amount:

(Name of Alternate):   Increase/decrease (circle one) Base Bid by:

(Numerical alternate amount) $

(Written alternate amount)   dollars

(Write additional alternates as required and number them.)

4. The Bidder will start and complete the work as follows.

Starting Date:

Date of Substantial Completion:

5. By this Bid Form submittal, the Bidder affirms having visited the project site and has full knowledge of existing conditions which affect the work.

6. By this Bid Form submittal, the Bidder affirms having reviewed all the Contract Documents including the following Addenda:

(*Note to specifier: List all Addenda provided to bidder.)

7. Submit bid qualifications and reasons with this Bid Form.
Bid qualifications may include: Proposed subcontractors, cash flow requirements, access to the work, assumptions for staging the work, assumptions for protecting existing and adjacent work, and proposed modifications to General and Supplementary Conditions.

8. Signed and sealed (Enter date, Bidder's signature and legal business address.)

List of Bid Qualifications by Bidder (if any):

END OF DOCUMENT

Notes:

___
___
___
___
___
___
___
___
___
___

00700
GENERAL CONDITIONS OF THE CONSTRUCTION AGREEMENT

*Note to specifier: If writing a set of GENERAL CONDITIONS for a project, you can use the following as a guide to content. These should not be used without reference to relevant AIA documents and a full understanding of the legal meaning and applicability of the text you use.

1.1 GENERAL

A. The Work includes all construction materials, labor, equipment, and services required by the Drawings, Specifications, and related Contract Documents. This includes labor, materials, etc. even if not explicitly required in the documents, as required to complete the project and provide project safety and security.

B. The project work shall comply with all applicable federal, state, county and/or city regulations. All building permits and other permits shall be obtained as required before starting work.

C. Substitution of any specified item is not permitted except through written request and written approval by the Architect. The Contractor's request must provide all specification data and certification that the substitution meets all requirements of the originally specified item. Any cost savings from an accepted substitution shall be shared equally by the contractor and the Owner.

2.1 OWNER'S RIGHTS AND RESPONSIBILITIES

A. The Owner shall provide a survey of the project site to the Contractor. The accuracy of the survey is not guaranteed by the Owner and the Contractor is responsible for verifying actual site dimensions and field conditions.

B. The Owner shall obtain and pay for all necessary approvals, easements and/or variances required for construction or preparation for construction.

C. If the Contractor fails to complete or correct any of the Work according to the Construction Documents the Owner may stop work until the work is completed or corrected.

D. The Owner reserves the right to occupy portions of the building space as work proceeds. If such occupancy interferes with the work, a negotiated fee or credit will be provided to the Contractor by the Owner and the required date of completion will be adjusted accordingly.

3.1 CONTRACTOR RIGHTS AND RESPONSIBILITIES

A. The Contractor shall have complete responsibility for, and control over, construction methods, techniques, procedures and project safety and security.

B. Any stated observations of possible safety or security hazards offered by the Architect or Owner in no way relieves the Contractor of full responsibility for such conditions.

C. The Contractor shall promptly pay for all labor, equipment, materials and services required to complete the Work as described in the Construction Agreement.

D. The contractor shall pay for special permits, inspections, tests, certifications, etc. required in the course of construction.

E. The Contractor shall be held responsible for all damages caused by his employees or subcontractors. The Contractor shall be held responsible for all errors, omissions, negligence, non-compliance with drawings and specifications, or uncorrected work by employees, suppliers, fabricators, and subcontractors.

F. The Contractor shall hold harmless the Owner from and against all claims, damages, losses, expenses, legal fees or other costs resulting from the Contractor's performance of the Work of the Construction Agreement.

G. The Contractor shall provide free access to the work by the Owner, Architect, and their representatives at all times.

3.2 SUBCONTRACTORS

A. The Contractor's choice of subcontractors, suppliers, and fabricators shall be as approved by the Architect and Owner.

B. The Contractor may not be required to use a specific subcontractor, fabricator, or supplier if there is an explicit reason not to that is confirmed by the Architect.

3.3 WORK BY OTHER CONTRACTORS

A. The Contractor and all subcontractors shall work in full cooperation on the project. This includes reasonable provision for storage and protection of Owner's, Contractor's, or subcontractor's materials and equipment.

4.1 CHANGES

A. Revisions, additions, or deletions to the Work under this agreement will be made by written order signed by Owner and Contractor. Such changes will not alter the remainder of this Agreement. If the revisions, additions, or deletions to the work affect the time and cost of project delivery, the Agreement will be amended accordingly.

5.1 TIME

A. If Work in this agreement is delayed by Owner-requested changes, delays in required approvals, labor disputes, fire, severe weather, or other conditions over which the Contractor has no control, the time for completion shall be extended accordingly and amended in writing in the Contract.

6.1 DISPUTES

A. Claims or disputes between the Contractor and the Owner arising from this Agreement that are not settled through negotiation shall be offered for mediation according to the rules of the American Arbitration Association. Disagreements not settled by mediation shall be offered for arbitration as per the rules of the American Arbitration Association. Work shall not be halted or slowed by the Contractor during negotiation, mediation, or arbitration of such disputes.

7.1 PAYMENTS

A. Payments will be made to the Contractor by the Owner according to the schedule in the Agreement.

B. Payments may be withheld because of:

Defective or non-conforming work not corrected.

Failure of the Contractor to make payments to subcontractors, suppliers, labor, or for services.

Failure to perform the Work according to the terms and conditions of this Agreement.

Legal or other claims by third parties related to the work of this Agreement.

C. Final payment shall be due when the Work is completed according to the Construction Documents and as stipulated in the Agreement between Owner and Contractor.

8.1 INSURANCE

A. The Contractor shall furnish the Owner the following certificates of insurance as previously noted in the Instructions to Bidders. The insurance amounts shall be as indicated or for larger amounts as required by law.

Workmen's Compensation Insurance for at least $500,000 each occurrence and $500,000 total for Bodily Injury including Personal Injury.

Property Damage for at least $100,000.

Comprehensive Automobile Liability for at least $250,000 for each person, $500,000 each occurrence, and Property Damage for at least $100,000 for each accident.

B. The Owner shall maintain property insurance in the full amount of insurable value. The insurance shall include the interests of the bank, or other mortgage holder, if any, and the Owner shall insure against "all risks" of physical loss or damage.

9.1 TERMINATION OF THE AGREEMENT

A. The Owner is required to provide prompt payments to the Contractor. If payment is not forthcoming, through no fault of the Contractor, the Contractor may terminate the contract, after seven (7) days written notice to the Owner.

10.1 JURISDICTION

A. This agreement shall be enforced according to the laws of the jurisdiction of the project.

This agreement entered into this ___ day of _____ , (year) _____ .

_____   _____
Owner' signature   Contractor's signature

_____   _____
Printed Name   Printed Name

END OF DOCUMENT

Notes:

_____
_____
_____
_____
_____
_____
_____
_____
_____
_____
_____
_____

Notes:

DIVISION 1

GENERAL REQUIREMENTS
01000

CONTENTS

01010	SUMMARY OF WORK	38
01020	ALLOWANCES	40
01022	TENANT ALLOWANCES	41
01025	SCHEDULE OF VALUES	42
01030	ALTERNATES	43
01045	CUTTING AND PATCHING	44
01100	CONSTRUCTION PROCEDURES	45
01153	CHANGE ORDER PROCEDURES	46
01300	SUBMITTALS	47
01500	TEMPORARY FACILITIES	48
01600	PRODUCTS AND SUBSTITUTIONS	49
01700	CONTRACT CLOSEOUT	50
01800	CLEANING AND MAINTENANCE	51

01010
SUMMARY OF WORK

GENERAL

1.1 SUMMARY

A. Project (*Note to specifier: add project name, client name, and location of project).

B. Project summary:

1. Project type and size.

*Note to specifier: Describe project as new, rehab, addition, restoration, etc. Describe building use, building height, number of floors, floors below grade, floor area, number of parking stalls and area of parking, number of rooms or units, etc.

2. Construction type and Occupancy type: (From building code.)

3. Project includes: (Architectural, Structural, Plumbing, Fire Protection, HVAC, Electrical.)

4. Performance requirements for completed work: (If not specified elsewhere.)

C. Project requirements: (A summary list of work included.)

Existing site conditions and restrictions:

Requirements for construction schedule, and sequence of work:

Previous or concurrent work by Owner or others:

Pre-purchased items:

Separate prime contracts:

Previous asbestos or hazardous waste abatement by Owner or others:

Items provided and/or installed by the Owner:

Items provided by Owner, installed by Contractor:

Early occupancy by Owner:

Occupancy of adjacent facilities:

Contractor's use of new and existing facilities by Contractor:

D. Apply, obtain, and pay for all permits required for the work. Submit two copies each of all permits to Owner and to Architect.

E. Comply with all applicable building codes and rules of other governing regulatory agencies. Submit two copies each of permits, inspection reports, and certificates of compliance to Owner and Architect.

F. Verify field dimensions before ordering fabrications or products to fit in place. Notify Architect of existing conditions and dimensions that differ from those shown in the Drawings.

G. Unless noted otherwise, the subject of all imperative sentences in the Specifications is the Contractor. For example, "Provide and install . . ." means "Contractor shall provide and install . . ."

END OF SECTION

Notes:

01020
ALLOWANCES

PART 1 -- GENERAL

1.1 SUMMARY

A. Allowances listed below are for materials only. Include in the base bid all other costs including labor and costs of additional adjacent or related construction.

B. Notify Architect in ample time when a decision on an allowance item is required to avoid a delay in construction.

C. Certify that quantities of products purchased are what are needed with reasonable allowance for waste and spare maintenance supplies for the Owner.

D. Submit invoices to show actual quantities and costs of materials delivered. Show applicable trade discounts.

PART 2 -- PRODUCTS

(Not applicable.)

PART 3 -- EXECUTION

3.1 SCHEDULE

A. Lump sum allowances:

1. Finish hardware: $

2. Signage: $

3. Interior landscaping: $

4. Exterior landscaping: $

END OF SECTION

01022
TENANT ALLOWANCES

*Note to specifier: This allowance is for construction of tenant improvements and typically lists quantities based on square footage for the following items:

Tenant entrance

Partitions

Ceilings

Flooring

Interior doors

Fixtures

Window treatments

Electrical supply

Lighting

Telecommunications

Sprinklers (as required by code, typically 1 per 150 sq. ft.)

Exit signs (as required by code, typically 1 per 3,000 sq. ft.)

HVAC

*Note to specifier: Delete those not applicable, add others as required.

END OF SECTION

Notes:

_____

_____

_____

_____

_____

01025
SCHEDULE OF VALUES

GENERAL

1.1 SUMMARY

A. Unless otherwise stated in the Agreement, provide a detailed breakdown of the Contract Sum as a Schedule of Values that are allocated to each part of the Work.

B. Before submitting the first application for payment, submit a proposed Schedule of Values to the Owner.

C. Provide copies of subcontracts and other data acceptable to the Owner to substantiate the sums described.

END OF SECTION

Notes:

01030
ALTERNATES

PART 1 -- GENERAL

1.1 SUMMARY

A. List the price for each alternate in the Bid Form. Include the cost of modifications to other work to accommodate each alternate. Include related costs such as overhead and profit.

B. The Owner will determine which alternates will be included in the Contract.

C. Alternates are listed in this section. See the Drawings and Specifications for particulars.

D. Coordinate alternates with related work to ensure that work affected by each selected alternate is properly executed.

PART 2 -- PRODUCTS

(Not applicable.)

PART 3 -- EXECUTION

*Note to specifier: Complete the following details of alternate items to be bid upon. Alternates should be kept to a minimum to avoid complications in the bidding process.

3.1 SCHEDULE

A. List of alternates:

Add alternates.

Deduct alternates.

Cost comparison of alternate.

Change in scope of work and schedule.

END OF SECTION

01045
CUTTING AND PATCHING

PART 1 -- GENERAL

1.1 SUMMARY

A. Cut and patch as required to complete the work for:

Visual quality as directed by the Architect.

Plumbing, HVAC, electrical, and communication systems.

Fire resistance ratings.

Inspection, preparation, and performance.

Cleaning. *Note to specifier: See Section 01800 on CLEANING.

B. Cut and patch with care to avoid damage to work, safety hazards, violation of warranty requirements, building code violations, or maintenance problems.

PART 2 -- MATERIALS AND PRODUCTS

2.1 MATERIALS

A. Match existing materials with new materials so that patching work is undetectable.

PART 3 -- EXECUTION

3.1 INSTALLATION

A. Inspect field conditions to identify all work required.

B. Notify Architect of work that might disrupt building operations.

C. Perform work with workmen skilled in the trades involved. Prepare sample area of each type of work for approval. Protect adjacent work from damage and dirt.

D. For cutting work, use proper cutting tools, not chopping tools. Make neat holes. Minimize damage to adjacent work. Check for concealed utilities and structure before cutting.

E. Make patches, seams, and joints durable and inconspicuous. Tolerances for patching shall be the same as for new work.

F. Clean work areas and areas affected by cutting and patching operations as described in Section 01800 on CLEANING.

END OF SECTION

01100
CONSTRUCTION PROCEDURES

PART 1 -- GENERAL

1.1 SUMMARY

A. Provide administrative coordination of all work, including trained, qualified employees and subcontractors, and supervisory personnel.

B. Arrange and conduct preconstruction and construction meetings with design principals, consultants, and construction trades when required by the Architect.

C. Submit progress schedule, bar-chart type, updated monthly. Provide submittal schedule, coordinated with progress schedule. Submit schedule of required tests including payment and responsibility.

D. Submit schedule of values.

E. Submit payment request procedures.

F. Provide to the Architect and post at the construction site, a phone and address list of individuals to be contacted in case of emergency.

G. Maintain and update record drawings and specifications as work progresses. Submit a complete, updated set of record documents upon conclusion of the work.

H. Keep all work clean and well protected from dirt, weather, theft, and damage.

END OF SECTION

Notes:

_____

_____

_____

_____

_____

_____

_____

01153
CHANGE ORDER PROCEDURES

PART 1 -- GENERAL

1.1 SUMMARY

A. Changes in the work may be required which will be authorized by a Change Order.

B. Change Orders, signed by the Owner and Architect, to authorize changes in the work will include equivalent changes in the Contract Sum and/or Time of Completion.

C. Change orders will be numbered in sequence and dated.

D. A request for estimates for possible changes is not a Change Order or a direction to proceed with the proposed changes. That can only be authorized through a signed Change Order.

END OF SECTION

Notes:

01300
SUBMITTALS

PART 1 -- GENERAL

1.1 SUMMARY

A. Provide all submittals as specified. Provide four copies where multiple copies are specified but the number is not stated.

B. Provide re-submittals when submittals are not approved.

C. Samples and shop drawings shall be prepared specifically for this project. Shop drawings shall include dimensions and details, including adjacent construction and related work. Note special coordination required. Note any deviations from requirements of the Contract Documents.

D. Provide warranties as specified. Warranties shall be signed by supplier or installer responsible for performance. Warranties shall not limit liability for negligence or non-compliance with documents.

*Note to Specifier: The following text is included in most sections in Quick Specs. On simpler building projects, it can be included here in General Requirements and cited by reference.

Submit the following within calendar days after receiving the Notice to Proceed.

*Note to specifier: Submittals are usually required within a specified number of calendar days after the Contractor is given the Notice to Proceed. Your choice of time will depend on the size of the project and the Owner's need for an expedited schedule.

Submit list of materials to be provided for this work.

Submit manufacturer's specifications required to prove compliance with these specifications.

Submit manufacturer's installation instructions.

Submit Shop Drawings as required with complete details and assembly instructions.

Submit Shop Drawings showing relationship and interface with adjacent or related work.

Submit samples of proposed exposed finishes and hardware for approval by the Architect.

END OF SECTION

01500
TEMPORARY FACILITIES

PART 1 -- GENERAL

1.1 REQUIREMENTS

A. Provide temporary services and utilities, including utility costs, for all services required for construction.

B. Provide construction facilities, including protected storage for building materials.

C. Provide construction access road and walkways as required.

D. Provide security and protection requirements including fire extinguishers as required by the local Fire Marshal, site enclosure fence, barricades, warning signs, security lighting, building enclosure, locking, security, and pest control.

E. Provide personnel support facilities including field office if required by the Architect, sanitary facilities, and drinking water.

F. Install and maintain project identification sign as designed and provided by the Architect.

END OF SECTION

Notes:

01600
PRODUCTS AND SUBSTITUTIONS

GENERAL

1.1 REQUIREMENTS

A. Provide products from one manufacturer for each type or kind as applicable. Provide secondary materials as recommended by manufacturers of primary materials.

B. Provide products selected or approved equal. Products submitted for substitution shall be submitted with acceptable documentation, and include costs of substitution including related work.

C. Conditions for substitution include:

An 'or equal' phrase in the specifications.
Specified material cannot be coordinated with other work. Specified material is not acceptable to authorities having jurisdiction. Substantial advantage is offered Owner in terms of cost, time, or other valuable consideration.

D. Substitutions shall be submitted prior to award of contract, unless otherwise acceptable. Approval of shop drawings, product data, or samples is not a substitution approval unless clearly presented as a substitution at the time of submittal.

END OF SECTION

Notes:

01700
CONTRACT CLOSEOUT

GENERAL

1.1 SUMMARY

A. The following are prerequisites to substantial completion. Provide the following:

Completed punch list and supporting documentation.

Signed warranties.

Certifications as specified.

Occupancy permit from governing agencies and utility companies as required.

Testing and start up of building systems.

Change and transfer of locks and keys as specified.

B. Provide the following prior to final acceptance:

Final payment request with supporting affidavits.

Completed punch list and supporting documentation.

C. Provide sets of record drawings showing original design and all changes made during construction.

D. Provide the following closeout procedures:

Submit record documents.

Submit maintenance manuals.

Complete all repairs, call-backs, corrections, re-adjustments of equipment, final cleaning, and final touch-up. Remove all temporary facilities, equipment, tools and supplies.

END OF SECTION

01800
CLEANING AND MAINTENANCE

GENERAL

1.1 SUMMARY

A. Keep the buildings and site well-organized and clean throughout the construction period.

B. Provide general clean up daily and complete weekly pickup and removal of all scrap and debris from the site. Exception: Reusable scrap shall be stored in a neatly maintained, designated storage area.

C. Weekly pickup shall include a thorough broom-clean sweep of all interior spaces. Also, each week, sweep paved areas on the site and public paved areas adjacent to the site. Completely remove swept dirt and debris. Daily and weekly cleanings will not replace required clean up after the work of specific trades such as specified herein.

D. At completion of the Work, remove from the job site all tools and equipment, surplus materials, equipment, scrap and debris.

E. Exterior of building: Inspect exterior surfaces and remove all waste materials, paint droppings, spots, stains or dirt.

F. Interior of building: Inspect interior surfaces and remove all waste materials, paint droppings, spots, stains or dirt.

G. Glass: Clean inside and outside so there are no spots or dirt, and no smudges or streaks remain from the cleaning process.

H. Schedule final cleaning as approved by the Owner to enable Owner to accept a completely clean Work.

I. Final cleaning will be comparable to that provided by professional, skilled cleaners using commercial grade cleaning materials. Cleaning materials will be used with care and will be compatible with building materials and finishes. Final cleaning will include removal of scraps or waste in landscaped areas and thorough cleaning of walkways, desks, paved areas and public paved areas adjacent to the site.

*Note to Specifier: The following text is included in most sections in Quick Specks. On simpler building projects, it can be included here in General Requirements and cited by reference.

After installation, inspect all work for improper installation or damage.

Operating hardware must perform smoothly. Repair or replace any defective work. Repair work will be undetectable. Redo repairs if work is still defective, as directed by the Architect. Clean the work area and remove all scrap and excess materials from the site.

END OF SECTION

Notes:

DIVISION 2

SITEWORK
02000

CONTENTS

02000	SITE PREPARATION	55
02050	DEMOLITION	57
02200	EARTHWORK	60
02500	PAVING	64
02510	CONCRETE ROADS AND WALKS	66
02520	ASPHALT PAVING	69
02600	UTILITIES	73
02600	UNDERGROUND UTILITIES	75
02660	WATER DISTRIBUTION	77
02700	SEWERAGE	79
02710	SUBDRAINAGE	81
02900	LANDSCAPING	84

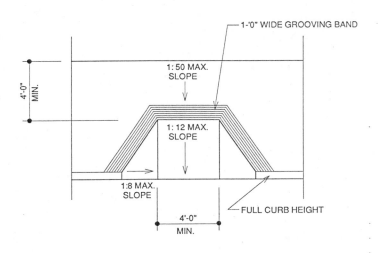

WHEEL CHAIR RAMP PLAN

DIVISION 2
SITEWORK

SAMPLE SITEWORK CONSTRUCTION DETAIL

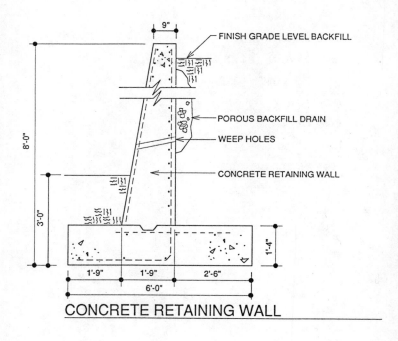

CONCRETE RETAINING WALL

DIVISION 2
SITEWORK

SITE PREPARATION, DEMOLITION, AND GRADING
02050 -- 02100

PART 1 -- GENERAL

*Note to specifier: This is generic introductory text that can be applied to the Demolition, Site Preparation, and Grading sections that follow.

1.1 WORK

A. Provide sitework as shown on the Drawings and specified herein.

B. Provide all related materials, equipment, and labor required to complete the work as specified.

C. Related work:

_ Grading and compaction as required.
  _ Excavation, trenching, and utilities.

1.2 QUALITY STANDARDS

A. Provide experienced, well-trained workers competent to complete the work as specified.

B. Unless approved by the Architect, provide all related products and accessories from one manufacturer.

C. Use materials:

_ From manufacturers and suppliers who specialize in paving materials.
_ From manufacturers and suppliers specified or approved by the Architect.

D. All work shall comply with governing building and safety codes.

1.3 SUBMITTALS

A. Submit the following within ___ calendar days after receiving the Notice to Proceed.

*Note to specifier: Submittals are usually required within a specified number of calendar days after the Contractor is given the Notice to Proceed. 30 calendar days is a common requirement for medium- to large-size projects. Your choice of time will depend on the size of the project and the Owner's need for an expedited schedule.

_ Submit list of materials to be provided for this work.
_ Submit manufacturer's specifications required to prove compliance with these specifications.
_ Submit manufacturer's installation instructions.

1.4 MATERIALS HANDLING

A. Provide all materials required to complete the work as shown on the Drawings and specified herein.

B. Deliver, store, and transport materials to avoid damage to the materials or to any other work.

1.5 PRECONSTRUCTION AND PREPARATION

A. Examine and verify that job conditions are satisfactory for speedy and acceptable work.

_ Maintain and use all up-to-date construction documents on site.
_ Maintain and use up-to-date trade standards and materials supplier's instructions.
_ Confirm there is no conflict between this work and governing building and safety codes.
_ Confirm there are no conflicts between this work and work of other trades.
_ Confirm that work of other trades that must precede this work has been completed.

B. Planning and coordination:

_ Notify Architect when work is scheduled to be started and completed.
_ If required by the Architect, a preconstruction meeting will be held with all concerned parties.
_ Use agreed schedule for installation and for field observation by Architect.

END OF SECTION

Notes:

_____
_____
_____
_____
_____
_____
_____
_____
_____

DIVISION 2
SITEWORK

DEMOLITION
02050

PART 1 -- GENERAL

*Note to specifier: Include generic introductory PART 1 text from the beginning of this Division.

PART 2 -- MATERIALS

2.1 PROTECTIVE BARRIERS AND COVERS

A. Provide demolition materials, barriers, protective covers, etc. to complete the work as specified.

PART 3 -- CONSTRUCTION

3.1 SITEWORK PREPARATION

A. Regulations and permits:

_ Obtain all required permits and approvals.
_ Obey all restrictions, deadlines, notification requirements, etc. of governing agencies.
_ Notify of impending sitework:
_ Owners of adjacent properties
_ Public agencies as required by code
_ Utility companies
_ Identify and clearly mark underground utility lines, pipe, cable, and conduits.

B. Coordination:

_ Keep and use the latest construction documents on site:
_ Architectural _ Civil _ HVAC pertaining to sitework
_ Plumbing pertaining to sitework
_ Electrical pertaining to sitework _ Landscaping

_ A preconstruction meeting will be held with all concerned parties, to review coordination.

3.2 SURVEY CHECK

A. Check site survey for errors, and make necessary corrections.

_ Verify locations and provide clear marking of:
_ Property lines _ Setback lines _ Easements _ Rights of way
_ Verify that site surveyor's marker locations are correct:
_ Monument _ Stakes _ Flags

3.3 SUBSURFACE INVESTIGATION/SOIL TESTS

A. Soil tests:

_ Clearly mark and maintain marks at soil test locations.
_ Conduct soil investigation and tests.
_ Keep soil reports on site.

B. Make sure that test pits, borings, and tests are adequate to find any potential soil problems.

_ Test pits and/or borings are as per Drawings in:
_ Number _ Locations _ Sizes _ Depths

_ Verify that soils testing engineer and laboratory are properly certified.
_ Verify elevation of water table.
_ Provide temporary drains or pumps as required to remove ground water.

3.4 SITEWORK -- DEMOLITION

A. Demolish and remove all work indicated on Drawings:

_ Follow all permit requirements and governing regulations.
_ Provide Owner with proof of compliance with permit requirements and governing regulations.
_ Verify demolition shown in Drawings in onsite review with the Architect.
_ Start demolition at topmost level, and proceed downward.
_ Provide water supply and hoses for spray, to control dust.
_ Provide operational plan and timeline for demolition.
_ Keep stairs, corridors, and entries/exits clear.

B. Protection of property:

_ Don't allow debris or dust to contaminate adjacent property.
_ Check and correct potential demolition hazards in existing construction.
_ Identify utilities for cutoff.
_ Locate any hidden utilities.
_ Remove all potentially dangerous or flammable materials.
_ Build and test the strength of protective bracing.
_ Provide braces or shores wherever structural elements will be removed in partial demolition.
_ Do not allow any dislodged materials to fall outside demolition area.
_ Protect all public areas and adjacent property with secure protective barriers.

C. Cleanup:

_ Provide debris chutes ample in size, thoroughly supported, and secure from movement.
_ Completely control and remove all demolition debris, scraps, and dust.

3.5 CONSTRUCTION PROTECTION

A. Provide sturdy barriers and covers as necessary for safety and to protect remaining work:

_ Pedestrian barriers
_ Shoring
_ Location markers for buried utilities
_ Covers over pits and trenches
_ Covers over paving, drives, and ramps
_ Tree and shrub protectors

B. Provide security lighting, fencing, and warning signs.

_ Protect property from intruders.
_ Provide warning lights to protect users of adjacent property.
_ Provide warning lights as necessary to protect those in public areas near this work.

3.6 PLANT PROTECTION

A. As directed by the Architect, label existing shrubs and trees:

_ To remain _ To be relocated _ To be removed

B. Erect barricades and fences as required to protect planting and related property.

C. Provide protection for existing and new plants:

_ Keep areas within the drip line of saved trees clear of:

_ Construction materials delivery or storage
_ Construction equipment
_ Traffic
_ Parking
_ Soil storage
_ Debris
_ Water ponding

_ Keep oil and other chemicals away from paving and planting.
_ Protect planting areas from foot and wheel traffic.
_ Do not allow excavation near trees to damage roots.
_ Do not prune trees except as approved by the Architect.
_ Pruning can only be done by qualified landscaping specialists.
_ Provide water and nutrients for existing plants that are to remain.
_ Obtain and follow the watering and nutrient instructions of the landscaping consultant.

END OF SECTION

DIVISION 2
SITEWORK

EARTHWORK -- EXCAVATION, GRADING, AND BACKFILL
02200

PART 1 -- GENERAL

*Note to specifier: Include generic introductory PART 1 text from the beginning of this Division.

PART 2 -- MATERIALS

2.1 FILL

A. Fill materials shall be:

_ Uniform   _ From an approved source
_ Sampled and tested as directed by the Architect.

PART 3 -- EXECUTION

3.1 PREPARATION

A. Protection:

_ Obtain and obey all applicable regulations regarding grading and excavation.
_ Identify, mark, and protect from damage all existing underground pipes, conduits, and cable:
_ Water supply   _ Sanitary sewer   _ Storm sewer   _ Gas
_ Steam   _ Electrical   _ Communications cable

_ Contact utility companies and public agencies to verify locations of underground services.
_ Keep overhead utility lines out of range of damage by trucks and cranes.
_ If utility lines are damaged, they shall be repaired or replaced by the Contractor.
_ Repair of damaged utilities shall be as directed by the Architect and utility company.
_ Existing utilities that will interfere with construction shall be safely relocated.
_ Relocation of utility lines shall be as directed by the Architect and utility company.
_ Provide amply engineered shoring and bracing as required by site conditions.
_ Provide temporary drains and/or pumps to remove ground water.

B. Prepare a survey record drawing to record all new site conditions.

3.2 GRADING AND EXCAVATION

A. Grade and excavate to lines, grades, and elevations as shown in the Drawings.

_ Keep grading for slabs:
_ Level _ Non-sloping
_ Not crowned toward center portion

_ Do not allow heavy machinery to operate above buried water, sewer, steam, or gas lines.

B. Topsoil control:

_ Establish topsoil depth.
_ Remove soil with organic matter, and store soil, if reusable for finish grading and landscaping.
_ Remove topsoil excluding:
_ Subsoil _ Rock _ Debris

3.3 EXCAVATION

A. Excavate for utilities, footings, and all other work shown in the Drawings and specified herein.

B. Excavation control:

_ Immediately investigate any unexpected subsurface conditions that appear during excavation:
_ Water _ Debris _ Rock fill, concrete, or bituminous fill

_ Depressions or soft spots
_ Abrupt change in soil type or density
_ Previous paving, footings, or other construction
_ Archaeological artifacts

_ Keep excavation pits and trench floors:
_ Level if drainage not required.
_ Consistent if sloped to drain.

_ Keep foundation and footing trenches uniform in width and direction as per Drawings.
_ Clean excavations of debris and loose dirt, and keep clean before pouring concrete.
_ Immediately remove dirt, rock, or other debris that spills onto paving or planting areas.
_ Take frequent measurements to prevent over-excavation.
_ Provide temporary drainage as necessary to prevent ponding, erosion, or spillover.

_ Inspect foundation excavations before pours for adherence to Drawings:
_ Trench drainage _ Footing drains _ Bracing
_ Clearances
_ Large boulders and rock to be removed will be removed at no additional cost to the Owner.

3.4 SITE MAINTENANCE DURING GRADING AND EXCAVATION

A. Control excavation dust:

_ With water spray
_ Through controlled demolition
_ No dust is to be allowed to blow onto neighboring property.

B. Cleaning:

_ Do frequent and thorough cleanups.
_ Identify potentially harmful substances that might be uncovered during excavation.
_ Handle potentially harmful substances strictly according to governing regulations.
_ Promptly remove them from site as per governing regulations.

3.5 BACKFILL AND COMPACTION:

A. Before backfilling:

_ Architect must approve work completed below finish grade.
_ Formwork at below grade construction must be removed.
_ Underground utilities that will be concealed must be inspected, tested, and approved.
_ Trash and debris must be removed.

B. Perform backfill and compaction in a systematic pattern, to assure complete and consistent work.

_ If any over-excavation accidentally occurs, correct it with well-compacted backfill.
_ Fill and thoroughly compact holes from root and stump removal pits.
_ Place required termite and other soil poisons along with backfilling.
_ Provide testing and inspection of backfill and compaction.
_ Layer backfill in 6 inch to 12 inch increments.
_ Compact all fill.
_ Use stabilized fill material of an approved type and from an approved source.
_ Test and approve fill material delivered from other sites.
_ Do not allow any debris to be mixed with fill.

C. Protect foundation and retaining walls during backfilling:

_ Cure concrete foundation or retaining walls to reach required strength before backfilling.
_ Brace foundation or retaining walls to prevent damage from backfilling.
_ Thoroughly waterproof basement foundation walls before backfilling.
_ Notify the Architect of below-grade waterproofing, and arrange for observation of the work.
_ Do not allow any damage to waterproofing from backfilling.
_ Alternately place backfill at two sides of a wall, to avoid unbalanced loading.
_ Correctly replace boundary markers, monuments, and stakes if they are moved or damaged.

3.6 SUBGRADE PREPARATION FOR PAVING

A. Identify and locate existing underground construction:

_ Vaults _ Manholes _ Drains _ Sewers _ Utility boxes

_ Utility mains:

_ Gas _ Water _ Steam _ Electrical

_ Provide graded slopes as required for:

_ Positive pavement slopes to drains _ Roadway crowns
_ Roadway turn banks

_ Backfill in layers, and thoroughly compact trenches or pits beneath paving.
_ Compact all fill.
_ Do even and systematic rolling and tamping so that all portions of grade are equally compacted.
_ Install base course firmly, and wet it down prior to concrete application.

_ Protect base course from:

_ Frost _ Flooding

3.7 SURFACE DRAINAGE

A. Provide drainage catchers for roof water as well as for surface runoff.

B. Provide surface storm drainage as per Drawings and free of impediments to smooth drain flow:

_ Continuous _ No narrow restrictions _ No barriers
_ No sharp changes in direction _ No sharp drops in grade
_ No level areas or depressions

3.8. IRRIGATION AND SPRINKLERS

A. Lay out the work as per Site Drawings.

_ Install pipe and materials as shown in Plumbing Drawings.

B. Electrical:

_ Do electrical work related to irrigation and sprinkler controls, and coordinate with plumbing.

C. Testing:

_ Pressure-test the system to make sure it's leak free.
_ Test-operate the system, and adjust until it operates correctly.

END OF SECTION

DIVISION 2
SITEWORK

PAVING
02500

PART 1 -- GENERAL

1.1 WORK

A. Provide and install paving materials as shown on the Drawings and specified herein.

B. Provide all related materials, equipment, and labor required to complete the work specified.

C. Other related work:

_ Grading and compaction as required.
_ Excavation, trenching, and utilities work required to be completed before paving.

1.2 QUALITY STANDARDS

A. Provide experienced, well-trained workers competent to complete the work as specified.

B. Unless approved by the Architect, provide all related products and accessories from one manufacturer.

C. Use materials:

_ From manufacturers and suppliers who specialize in paving materials.
_ From manufacturer and suppliers specified or approved by the Architect.

D. All work shall comply with governing building and safety codes.

1.3 SUBMITTALS

A. Submit the following within _____ calendar days after receiving the Notice to Proceed.

*Note to specifier: Submittals are usually required within a specified number of calendar days after the Contractor is given the Notice to Proceed. 30 calendar days is a common requirement for medium- to large-size projects. Your choice of time will depend on the size of the project and the Owner's need for an expedited schedule.

_ Submit list of materials to be provided for this work.
_ Submit manufacturer's specifications required to prove
  compliance with these Specifications.
_ Submit manufacturer's installation instructions.

1.4 MATERIALS HANDLING

A. Provide all materials required to complete the work as shown on the Drawings and specified herein.

_ Deliver, store, and transport materials to avoid damage to the product or to any other work.
_ Return any products or materials delivered in an unsatisfactory condition.
_ Materials and products delivered will be certified by the manufacturer to be as specified.

B. Store materials in a safe, secure location, protected from weather.

1.5 PRECONSTRUCTION AND PREPARATION

A. Examine and verify that job conditions are satisfactory for speedy and acceptable work.

_ Maintain and use all up-to-date construction documents on site.
_ Maintain and use up-to-date trade standards and materials supplier's instructions.
_ Confirm there is no conflict between this work and governing building and safety codes.
_ Confirm there are no conflicts between this work and work of other trades.
_ Confirm that work of other trades that must precede this work has been completed.

B. Planning and coordination:

_ Notify Architect when work is scheduled to be started and completed.
_ If required by the Architect, a preconstruction meeting will be held with all concerned parties.
_ Use agreed schedule for installation and for field observation by the Architect.

END OF SECTION

Notes:

_____

_____

_____

_____

_____

DIVISION 2
SITEWORK

CONCRETE ROADS AND WALKS
02510

PART 1 -- GENERAL

*Note to specifier: Include generic introductory PART 1 text from the beginning of Division 2.

PART 2 -- MATERIALS

2.1 CONCRETE PAVING

A. Concrete:

_ All mixing and tests to assure compliance with standards as per ACI 301.
_ Provide concrete ready-mixed in compliance with ASTM C 94.
_ On-site mixed concrete will conform to ASTM C 685.
_ Concrete strength will conform to ACI 301, 318, and applicable building code requirements.

_ Compressive strength of   psi in 7 day test.
_ Compressive strength of   psi in 28 day test.
_ Slump   to   inches.
_ Minimum water cement ratio:

*Note to specifier: Building code and engineering requirements must be consulted to establish the standards of concrete strength listed above.

_ Add air-entraining admixture as required to protect concrete exposed to exterior weather.
_ Admixture proportions as per ACI 301, ACI 318, and manufacturer's instructions.

B. Forms:

_ Provide metal or wood formwork for borders and curbs with profiles to match detail Drawings.
_ Earth forms are not allowed for paving.

C. Reinforcing:

_ Comply with ACI 301 and related ACI, CRSI, and ASTM standards.
_ Deformed bars for number 3 and larger unless shown otherwise on Drawings.
_ No. 10 welded wire mesh, plain type in coiled rolls, unfinished.

D. Aggregate:

_ sub-base aggregate to depth shown in Drawings.
_ Maximum size is 3/4", compacted to 95%.

PART 3 -- CONSTRUCTION AND INSTALLATION

3.1 PREPARATION

A. Work conditions:

_ Examine site conditions and correct any conditions detrimental to the work.
_ Acquire and follow rules of all governing agencies with jurisdiction over the work.
_ Record dates and times of placement, interruptions, tests, completion, and finish work.
_ Establish a system for recording batching, mixing, placing, and curing.
_ Do not do work when new paving might be harmed by rain or low temperatures.
_ Verify that all necessary subgrade preparation is completed.

_ Install related work before concrete pour, and protect from damage:

_ Formwork   _ Anchors for site furniture and fixtures
_ Base plates   _ Inserts _ Bolts
_ Sleeves for bollards and fence posts   _ Utility boxes
_ Drains _ Electrical conduit   _ Pipe and plumbing
_ Separation joints   _ Headers/screeds

_ Keep pour area free of scraps, trash, and organic matter.
_ Secure screed boards against displacement during pour.
_ Install screed boards at correct height for paving thickness.
_ Use redwood or preservative-treated wood for screeds, border boards, and joint boards.
_ Brace formwork to maintain work at the lines and grades shown on the Drawings.

3.2 CONCRETE MIX AND PLACEMENT:

A. Place concrete according to ACI 301.

_ Verify that plant mix is certified for proportions.
_ Don't allow trucks to wait beyond time limits before pour.
_ Don't allow unauthorized watering; do not over-water.
_ Don't permit segregation.
_ Verify that visual slump is correct.
_ Do compaction, consolidation, and vibration as required.
_ Conduct tests as required, and submit reports.

B. _ Construct reinforcing as detailed:

_ Spacing _ Ties _ Laps _ Chairs/stirrups/supports
_ Vertical positioning within slab
_ Clearance to allow concrete flow   _ Free of loose scale
_ Clean of dirt or grease
_ Secure against dislocation during concrete pour.
_ Don't allow reinforcing to be in contact with dissimilar metals.
_ Avoid small or angular concrete paving sections, or install extra reinforcing to prevent cracking.
_ Keep pour area free of scraps, trash, and organic matter.

C. Joints:

_ Provide movement and relief joints in locations, depths, and widths as detailed:
_ At contact of pavement with other work.
_ For thermal expansion/contraction.
_ To control movement and settlement cracks.
_ At breaks in the construction sequence.

_ Make joint lines straight and uniform:
_ Coordinate and align with other work.
_ Prevent damage to joint lines.
_ Repair joint lines that are accidentally damaged.

_ Coordinate and align sawn joint work with other work.

D. Finishing and curing:

_ Floating, troweling, and texture finishes:
_ Do not trowel until bleed water is gone.
_ Do not over-trowel.
_ Do not apply dust to cement to speed up troweling start time.

_ Use approved coverings and curing and wetting methods.

E. Protection and repair or replacement:
_ Repair or replace defective work.
_ Protect fresh pavement from foot or traffic damage.

END OF SECTION

Notes:

___
___
___
___
___
___
___
___
___

DIVISION 2 -- SITEWORK

ASPHALT PAVING
02520

PART 1 -- GENERAL

*Note to specifier: Include generic introductory PART 1 text from the beginning of Division 2.

PART 2 -- MATERIALS

2.1 AGGREGATES AND ASPHALT

A. Sub-base and base aggregates:

_ Sub-base aggregate maximum size, 1-1/2".

_ Base aggregate maximum size:

_ Base courses over 6" thick, 1-1/2".
_ Other base courses: 3/4".
_ Compacted to 95%.

B. Asphalt:

_ Comply with Asphalt Institute Specification SS-2:

_ Asphalt cement penetration grade 50/60.
_ Prime coat cut-back type, grade MC-250.
_ Tack coat uniformly emulsified, grade SS-1H.

_ Hot plant mixed asphaltic concrete paving materials:

_ Temperature of material leaving the plant:

_ 290 degrees F. minimum
_ 320 degrees F. maximum

_ Temperature when placed: 280 degrees F. minimum.
_ Sealer as per Asphalt Institute Specification SS-2.

C. Paving borders:

_ Construction grade Redwood as required to maintain solid paving
    borders.

PART 3 -- CONSTRUCTION AND INSTALLATION

3.1 PREPARATION

A. Work conditions:

_ Do not do work when paving might be harmed by rain or low temperatures.
_ Verify that all necessary subgrade preparation is completed.
_ After grading and preparation of subgrade:
_ Scarify and sprinkle area to be paved.
_ Compact to hard, smooth surface of 90% compaction.
_ Apply weed killer according to manufacturer's instructions.

_ Install headers and stakes to define boundaries and paving pattern on Drawings.
_ At correct height for specified paving thickness.
_ Secure against movement.

_ Remove all loose materials from area of compacted base.

_ Do not place asphaltic concrete during inclement weather:

_ Outdoor temperature is below 50 degrees F.
_ During rain, fog, snow, or potential storm conditions.

_ Before adding finish paving:

_ Add a new base course over any aggregate used for temporary construction roads.

3.2 APPLICATION

A. Record dates and times of placement, interruptions, tests, completion, and finish work.

B. Base course application:

_ Uniformly apply base course materials of correct thickness.
_ Place base courses:

_ Sub-base:

_ Spread sub-base material to total compacted thickness shown in Drawings.
_ Compact to 95%.

*Note to specifier: sub-base is for larger paving areas and special conditions. Verify need for sub-base with civil engineer.

_ Base:

_ Spread sub-base material to provide total compacted thickness shown in Drawings.
_ Compact to 95%.

C. Asphalt application:

_ Use approved bituminous material mix.
_ Verify that bituminous material has been tested and certified at the plant.
_ Keep delivered and applied bituminous mix temperatures under control.
_ Cover asphalt with tarp to maintain required temperature until unloaded.
_ Spread material in the most direct manner to minimize handling.
_ Spread in one layer if finish paving is 3" or less.
_ Fit and shape paving closely around:
_ Catch basins _ Manholes _ Meter boxes
_ Properly slope paving surface to drain.

D. Rolling:

_ Roll out irregularities such as humps or dips.
_ Operate rollers only where soil has been compacted.
_ Do not allow rollers to damage curbs.
_ Do not allow rollers to damage adjacent paving.
_ After spreading, roll until surface is hard and smooth to finish elevations shown on Drawings.
_ Roll in multiple directions until no roller marks are visible.
_ Smoothness tolerance: 1/8" in 6'.

E. Thickness tolerance:

_ Confirm final bituminous paving thickness at randomly selected locations.
_ Compacted thickness within tolerance of minus 0.0" to plus 0.5".
_ Tolerance limit 3/8" in 10' maximum deviation from grade elevations shown on Drawings.
_ Moisture content shall be only as required for specified compaction.
_ Variation from true elevation may be 1/2" or less.

F. Final cleaning, tests, and repairs:

_ Redo and recompact the work as required to meet specified standards.
_ Verify required bituminous mix temperature after final rolling.
_ Remove spillover material, and clean adjacent pavements of spills.
_ Repair any damage to adjacent areas.
_ Repair deviations from smoothness tolerance limit by removing and replacing materials.

G. Flood test:

_ Before sealing, flood test pavement in presence of Architect.
_ Thoroughly flood paved area with water.
_ If water ponds to depth of more than 1/8" in 6', fill and correct to provide drainage.
_ Smooth the edges of all corrected work until repairs are invisible.

H. Seal coat:

_ Apply according to manufacturer's instructions.
_ Apply to create a finish seal which drys to a uniformly smooth, black, surface.
_ Protect sealed area from traffic until sealer has cured and is safe for traffic.

END OF SECTION

Notes:

DIVISION 2
SITEWORK

UTILITIES
02600

PART 1 -- GENERAL

1.1 WORK

A. Provide:

*Note to specifier: Name the specified item, such as Gas Supply Line, Water Supply Line, Sewer, etc.

B. Provide everything required to complete the Work as shown on the Drawings and specified herein.

C. Other related work:

*Note to specifier: This would include reference to related construction that is not part of this Section such as adjacent excavation, grading, paving, etc.

1.2 QUALITY STANDARDS

A. Provide experienced, well-trained workers competent to complete the Work as specified.

B. Unless approved by the Architect, provide all related products and accessories from one manufacturer.

C. Use products and accessories:

_ From a manufacturer who specializes in manufacturing products of this type.
_ From a manufacturer specified or approved by the Architect.

D. All work shall comply with manufacturer's instructions and governing building and safety codes.

1.3 SUBMITTALS

A. Submit the following within _____ calendar days after receiving the Notice to Proceed.

*Note to specifier: Submittals are usually required within a specified number of calendar days after the Contractor is given the Notice to Proceed. 30 calendar days is a common requirement for medium- to large-size projects. Your choice of time will depend on the size of the project and the Owner's need for an expedited schedule.

_ Submit list of materials to be provided for this work.
_ Submit manufacturer's specifications required to prove compliance with these Specifications.
_ Submit manufacturer's installation instructions.
_ Submit Shop Drawings if required for special installations.

1.4 MATERIALS HANDLING

A. Provide all materials required to complete the Work as shown on Drawings and specified herein.

_ Deliver, store, and transport materials to avoid damage to the product or to any other work.
_ Return any products or materials delivered in a damaged or unsatisfactory condition.
_ Materials and products delivered will be certified by the manufacturer to be as specified.

B. Store materials in a safe, secure location, protected from dirt, moisture, contaminants, and weather.

1.5 PRECONSTRUCTION AND PREPARATION

A. Examine and verify that job conditions are satisfactory for speedy and acceptable work.

_ Maintain and use all up-to-date construction documents on site.
_ Maintain and use up-to-date trade standards and manufacturer's instructions.
_ Confirm there is no conflict between this work and governing building and safety codes.
_ Confirm there are no conflicts between this work and work of other trades.
_ Confirm that work of other trades that must precede this work has been completed.
_ Meet all requirements to secure any applicable warranty.

B. Planning and coordination:

_ Notify Architect when work is scheduled to be started and completed.
_ If required by the Architect, a preconstruction meeting will be held with all concerned parties.
_ Use agreed schedule for installation and for field observation by the Architect.

END OF SECTION

Notes:

_____

_____

_____

_____

_____

_____

DIVISION 2
SITEWORK

UNDERGROUND UTILITIES
02600

PART 1 -- GENERAL

*Note to specifier: Include generic introductory PART 1 text from the beginning of this Division.

PART 2 -- MATERIALS

2.1 *Note to specifier: See specific utilities work in the pages that follow.

PART 3 -- CONSTRUCTION -- UNDERGROUND UTILITIES

3.1 PREPARATION AND COORDINATION

A. Identify work to be completed:

  _ By project subcontractors.
  _ By local utility companies.
  _ Any special work to be provided by client.

B. Coordinate utility trenching:

  _ Keep open trenchwork to a minimum.
  _ Avoid duplicate trenching.
  _ Avoid potentially damaging crossover trenching.
  _ Install sanitary sewer drainage first, followed by:
     _ Storm sewers _ Water supply _ Gas
     _ Electrical and communications

C. Check utility company or subcontractor plans to assure their work is coordinated with:

  _ Other work.
  _ Existing trees and planting that are to be protected.
  _ Existing construction.

3.2 INSTALLATION -- GENERAL INSTRUCTIONS FOR ALL UTILITIES

A. Install pipe and cable work as per instructions of manufacturers and utility companies.

  _ Delivery and storage protection _ Laying _ Jointing
  _ Trench support _ Protective covers
  _ Repair of coating or wrapping _ Cathodic protection
  _ Tests _ Backfilling _ Compaction

B. Safety and protection:

_ Protect trenches against bulging soil and cave-ins.
_ Seal openings of uncompleted pipework closed during off-work hours.
_ Install coated or wrapped pipe with care to avoid damage.
_ Repair any damaged portions of coated or wrapped pipe to match the protection of the original.
_ Determine that anticorrosion coating on steel piping is unbroken.
_ Provide thorough caulking of pipe thread connections.
_ Provide electrical conducting wire or other cathodic protection where metal piping includes non-conductive, flexible joints.

C. Valve and meter boxes:

_ Place valve and meter boxes level with final elevation of finish pavement or landscaping.
_ Place valve and meter boxes on compacted soil.
_ Do not allow any part of valve and meter boxes to rest on piping or valves.

D. Installation:

_ Verify connection point to building service lines.
_ Excavation and pressure tests must comply with utility company requirements and codes.
_ The contractor shall take every possible precaution to assure gas line safety.
_ Install shutoff valves as detailed in type and location.
_ Slope gas lines to main lines, and slope main lines to drip catchers, to avoid water accumulation.
_ Do not allow gas lines to make contact with other piping or conduit.
_ Keep gas lines beyond code minimum distance from water lines.

E. Protection of pipe and excavation backfill:

_ Protect pipe from freezing.
_ Pipe depth to comply with utility company requirements and local regulations.
_ Do not permit heavy equipment traffic over or near gas line trenches.
_ Clearly mark backfilled gas line trenches as warning to other excavators.

F. Completion:

_ Test as required by governing agency and utility company:

_ Connect to utility line or meter as per utility company requirements.
_ Repair or replace defective work as directed by the Architect.

END OF SECTION

DIVISION 2
SITEWORK

WATER DISTRIBUTION
02660

PART 1 -- GENERAL

A. Provide water distribution as shown on the Drawings and specified herein.

*Note to specifier: Include generic introductory PART 1 text from the beginning of this Division.

PART 2 -- MATERIALS

*Note to specifier: These suggested materials are suited to residential and small commercial construction. For larger projects, use a professional engineering consultant and rely on the consultant's specifications.

2.1 Water piping, below and outside building:

A. Pipe and fittings:

_ Cast Iron Pipe as per ANSI/AWWA C151
_ Ductile iron fittings.
_ Rubber gaskets.
_ Mechanical joints and 3/4" diameter rods.
_ Copper Tubing as per ASTM B88, Type K annealed.
_ Wrought copper fittings and compression joints.
_ PVC Pipe as per ASTM D1785, Schedule 40, or ASTM D2241.
_ PVC pipe shall resist a minimum of 150 psi with solvent weld joints.
_ Fittings and specials shall resist 150 psi unless specified otherwise.

B. Plastic pipe:

_ Fittings and specials as per Schedule 40 rating unless specified otherwise.
_ Threaded PVC fittings as per Schedule 80.

C. Valves:

   _ Working pressure of no less than 150 psi

PART 3 -- CONSTRUCTION AND INSTALLATION

3.1 INSTALLATION

A. Pipe hookup and protection:

_ Verify connection point to building service lines.
_ Provide every means necessary to prevent pipe movement from momentum of water flow.
_ Provide all necessary ties, anchors, and thrust blocks at turns, closed ends, and at large outlets.
_ Do not install water mains in same trenches with storm or sanitary sewers.
_ Observe a minimum distance of 10 feet between water mains and storm or sanitary sewers.
_ Protect pipe from freezing temperatures.

_ Test final system as required by governing building code:
  _ Sterilization _ Water pressure

B. Completion:

_ Test as required by governing agency and utility company:
_ Connect to utility line or meter as per utility company requirements.
_ Repair or replace defective work as directed by the Architect.

END OF SECTION

Notes:

_____
_____
_____
_____
_____
_____
_____
_____
_____
_____
_____
_____
_____

DIVISION 2
SITEWORK

SEWERAGE
02700

PART 1 -- GENERAL

1.1 WORK

A. Provide sanitary sewerage as shown on the Drawings and specified herein.

*Note to specifier: Include generic introductory PART 1 text from the beginning of this Division.

PART 2 -- MATERIALS

*Note to specifier: These suggested materials are suited to residential and small commercial construction. For larger projects, use a professional engineering consultant and rely on the consultant's specifications.

2.1 PIPE AND FITTINGS

A. Pipe:

_ Cast iron soil pipe and fittings (CIP): Class SV.
_ Rubber gaskets (ASTM C564) for compression joints.

B. Clay pipe and fittings (VCP):

_ Extra strength bell and spigot.
_ Use compression joints, type II.

C. PVC pipe and fittings:

_ Extra strength, minimum of SDR 35.
_ ABS pipe and fittings: ASTM D2680.

PART 3 -- INSTALLATION

3.1 PIPE LAYING

A. Sewer line must be placed exactly according to the approved drawing:

_ Protect pipe from damage.
_ Do not allow pipe to fall during placement.
_ Remove any extraneous material from pipe interior.
_ Lay pipe from lower grade to upgrade.
_ Lay with spigot ends of bell and spigot pipe aligned in direction of flow.
_ Install and adjust joints to assure watertightness.
_ Connect at building where shown on the Drawings.
_ Provide temporary plugs where pipe is not completed.
_ Place markers at grade ends of plugged lines.

B. Sanitary sewer lines and manholes:

_ Locate manholes so that covers will align with finish elevations of grade or landscaping.
_ Make firm bedding; compact if necessary, before laying pipe.
_ Align slopes from manhole to manhole.
_ Flush sewer lines clear of dirt and debris.
_ Provide manhole inverts as detailed.
_ Install cleanouts:
  _ Accessible for maintenance   _ Capped

C. Cleanup and repair:

  _ Remove all debris and excess materials from site.
  _ Replace or repair defective work as directed by the Architect.

END OF SECTION

Notes:

DIVISION 2
SITEWORK

SUBDRAINAGE SYSTEMS
02710

PART 1 -- GENERAL

1.1 SUBGRADE DRAINAGE

A. Provide subgrade drainage:

　　_ At building exterior perimeter
　　_ At each side of dividing fire wall foundation
　　_ Around elevator pit foundation walls
　　_ Drain to collection pit at designated corner of crawl space.

B. Basement Floor Slab Drainage:

　　_ Underside of slab, in rows 6 feet on center as indicated on Drawings.
　　_ Slope to drain at 1/4 inch per foot to sump pit.

PART 1 -- GENERAL

1.1 WORK

A. Provide building perimeter and underfloor weep drainage system, with filter aggregate.

　　_ Provide all piping, related work, and hookups as shown on the Drawings and specified herein.

*Note to specifier: Include generic introductory PART 1 text from the beginning of this Division.

PART 2 -- MATERIALS

2.1 PIPE

A. Provide drainage pipe as follows:

PIPE MATERIAL SCHEDULE

Pipe	Type	Manufacturer
1.		
2.		
3.		

B. Polyvinyl Chloride Pipe:

_ ASTM D2729, perforated, 6" inside diameter with required fittings.

*Note to specifier: Unperforated pipe may be selected or others as follow depending on site conditions and local custom.

C. Corrugated Plastic Tubing:

_ Flexible type, perforated, 4" diameter with required fittings.

2.2 FILTER AGGREGATE

A. Coarse Filter Aggregate:

_ Clean, well graded gravel or crushed stone
_ Free from clay, shale, organic material or debris.

B. Fine Filter Aggregate:

_ Clean, natural sand.
_ Free from clay, shale, organic material or debris.

2.3 ACCESSORIES

A. Joint cover

_ Joint Cover: No. 30 asphalt saturated roofing felt. 10 mil polyethylene.

*Note to specifier -- options:

_ Pipe Sleeve: Continuous or perforated plastic.
_ Filter Fabric:
_ Water pervious type, black polyolefin or polyester.
_ Manufactured by:

PART 3 -- CONSTRUCTION AND INSTALLATION

3.1 PREPARATION

A. Verify that excavation is ready to receive work.

*Note to specifier: Aggregate for drainage depends on site soil types, percolation, compaction and soils engineering requirements.

_ Compensate for over excavation with filter aggregate.

3.2 INSTALLATION

A. Install pipe and fittings as per manufacturer's instructions.

*Note to specifier: The following two options are for loosely butted and mechanically joined perforated pipe respectively.

_ Loose butt pipe ends:

_ Place 12" wide joint cover strip completely around pipe
_ Centered cover strip over joint.

OR

_ Place pipe with perforations downwards.
_ Attach pipe joints as per manufacturer's instructions.
_ Place drainage tile on bed of filter aggregate as shown on Drawings.
_ Lay pipe according to slope shown on Drawings.
_ Maximum variation from true slope of 1/4" in 10'

B. Place aggregate:

_ Apply coarse aggregate at sides and top of pipe.
_ Apply aggregate to 12" compacted depth at top of pipe.
_ Level top surface of aggregate cover.
_ Place filter fabric over aggregate cover before backfilling.
_ Place aggregate in maximum 4" lifts.
_ Consolidate each lift.
_ Increase compaction of each successive lift.
_ Avoid moving or damaging pipe during compaction.

C. Connection:

_ Connect to storm sewer system or sump pits with unperforated pipe couplings.

D. Cleanup and repair:

_ Remove all debris and excess materials from site.
_ Replace or repair defective work as directed by the Architect.

END OF SECTION

Notes:

_____
_____
_____
_____
_____
_____
_____
_____
_____
_____

DIVISION 2
SITEWORK

LANDSCAPING
02900

PART 1 -- GENERAL

1.1 WORK

A. Provide and install trees, plants, and ground cover as shown on the Drawings and specified herein.

B. Provide all related materials, equipment, and labor required to complete the Work as specified.

C. Other related work:

_ Grading and compaction as required.
_ Excavation, trenching, and utilities work required to be completed before planting.

1.2 QUALITY STANDARDS

A. Provide experienced, well-trained workers competent to complete the Work as specified.

B. Unless approved by the Architect, provide all materials from one supplier.

C. Use products and accessories:

_ From a supplier who specializes in the specified landscape materials.
_ From a supplier specified or approved by the Architect.

D. All work shall comply with governing building, safety, and zoning codes.

1.3 SUBMITTALS

A. Submit the following within _____ calendar days after receiving the Notice to Proceed.

*Note to specifier: Submittals are usually required within a specified number of calendar days after the Contractor is given the Notice to Proceed. 30 calendar days is a common requirement for medium- to large-size projects. Your choice of time will depend on the size of the project and the Owner's need for an expedited schedule.

_ Submit list of materials to be provided for this work.
_ Submit supplier's specifications required to prove compliance with these Specifications.
_ Submit supplier's planting instructions.

*Note to specifier: Samples may be requested with a different time limit than other submittals.

1.4 MATERIALS HANDLING

A. Provide all materials required to complete the Work as shown on Drawings and specified herein.

_ Deliver, store, and transport materials to avoid damage to the product or to any other work.
_ Return any materials delivered in an unsatisfactory condition.
_ Materials delivered will be certified by the supplier to be as specified.

B. Store materials in a safe, secure location, protected from weather.

1.5 PRECONSTRUCTION AND PREPARATION

A. Examine and verify that job conditions are satisfactory for speedy and acceptable work.

_ Maintain and use all up-to-date construction documents on site.
_ Maintain and use up-to-date trade standards and materials supplier's instructions.
_ Confirm there is no conflict between this work and governing building and safety codes.
_ Confirm there are no conflicts between this work and work of other trades.
_ Confirm that work of other trades that must precede this work has been completed.

B. Planning and coordination:

_ Notify the Architect when work is scheduled to be started and completed.
_ If required by the Architect, a preconstruction meeting will be held with all concerned parties.
_ Use agreed schedule for installation and for field observation by the Architect.

PART 2 -- MATERIALS

2.1 PLANTS, SOD, AND RELATED MATERIALS

A. Provide plants and related materials listed below:

_ Plant materials must be from a fully qualified plant supply nursery approved by the Architect.
_ Plant materials shall be certified by independent inspection.

PLANT SCHEDULE

Species	Quantity	Size	Location

B. Sod:

_ Provide general ASPA approved grade cultivated grass sod.
_ Strong fibrous root system.
_ Machine cut with 1/2 inch to 1 inch topsoil base.

_ Grass type must be suitable to local climate, microclimate, and growing conditions.

C. Fertilizer:

*Note to specifier: Name the manufacture, product trade name, and any related data required to clearly identify the intended product.

PART 3 -- LANDSCAPE INSTALLATION

3.1 COORDINATION AND PREPARATION

A. Coordination:

_ Coordinate planting with site improvements not yet installed:

_ Drains _ Irrigation _ Site furniture _ Paving _ Walls and fences

_ Prepare planting beds and pits as shown on plans.
_ Drain planting pits well, and keep them free of standing water.
_ Prepare planting pits of ample size to hold roots and root balls.
_ Permit finish grade to drain without interruption or diversion due to construction.

_ Provide topsoil as per instructions of the plant supplier:

_ Workability _ Consistency
_ Topsoil shall be clean of:
_ Foreign matter _ Clays _ Gravel _ Subsoil

_ See subsoil grading elevations for thickness of topsoil layer.
_ Till and loosen subsoil, to bond with topsoil.
_ Do all required subsoil and topsoil bond preparation.
_ Keep subsoil free of foreign matter and construction debris.
_ Compact topsoil evenly.

3.2 PLANTING PROCEDURES

A. Prepare soil, provide water, and install plants according to the instructions of the plant supplier.

B. Application:

_ "Water in" plants during planting.
_ Prune plants as required for renewed growth.

C. Protection:

_ Protect tree trunks after planting:
_ Wraps _ Wood and wire supports and braces
_ Vandal barriers _ Mulching

_ Protect planting areas from potentially damaging work such as:

_ Foot or machine traffic.
_ Tar kettles.
_ Concrete etching chemicals.
_ Plaster.
_ Cleaning materials, solvents, and oils.
_ Sand blasting.

_ Reject and replace all damaged plant materials:

_ After shipping and handling
_ After exposure to wind or sun
_ After repairs
_ After planting

_ Expedite transplanting and seeding to avoid plant damage.

_ Handle plants with care and as instructed by the supplier:
_ Don't lift earth ball plants by trunks or stems.
_ Don't lift container plants by trunks or stems.

D. Warranty and replacement:

_ Provide one year warranty to replace plants that die regardless of proper maintenance by Owner.

_ Replacement plants shall be of the same size and species as specified.
_ Plant replacements in the next growing season
_ Provide warranty for new plants commencing from date of replacement.

END OF SECTION

Notes:

_____

_____

_____

_____

_____

_____

DIVISION 2
SITEWORK

SAMPLE LANDSCAPE CONSTRUCTION

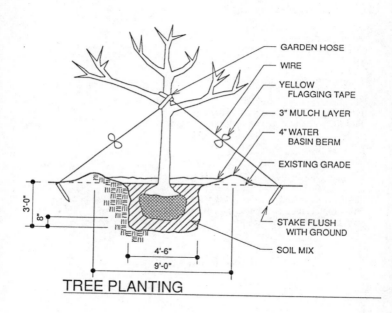

TREE PLANTING

DIVISION 3

CONCRETE
03000

CONTENTS

03300	CAST-IN-PLACE CONCRETE	90
03370	CURING AND FINISHING	103

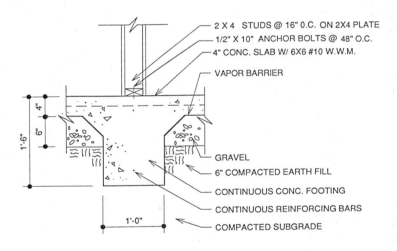

CONCRETE FOOTING & SLAB 1 Story

DIVISION 3
CONCRETE

CAST-IN-PLACE CONCRETE
03000

PART 1 -- GENERAL

1.1 WORK

A. Provide and install:

_ Formwork   _ Reinforcing   _ Formwork accessories
_ Cast-in-place concrete

1.2 SUBMITTALS

*Note to specifier: Submittals may not be required if A/E Drawings provide sufficient descriptive details.

A. Shop Drawings, where required by the Architect, will include:

_ Complete dimensions.
_ Required accessories such as joints, ties, bracing.
_ Reinforcing placement instructions as per applicable ACI and CRSI
   standards.
_ Reinforcement types and sizes and spacing
_ Instructions, and standards for fabrication as per applicable ACI and
   CRSI standards.
_ Schedule for erection and removal of formwork and accessories.
_ Product data including:
_ Admixtures.
_ Formwork materials.
_ Related accessories including all required ties, anchors, braces, inserts, etc.

_ Structurally engineered formwork will be as per:
   _ Drawings and calculations of a licensed structural engineer
      approved by the Architect.
   _ Restrictions of applicable building and safety codes.

1.3 QUALITY CONTROL, STANDARDS AND TOLERANCES

*Note to specifier: See REFERENCES on the CONTENTS page for this Division for design standards which are made a part of these Specifications by reference.

A. Place concrete according to ACI 301.

_ Reinforcing to comply with ACI 301 and related ACI, CRSI, and ASTM
   standards.
_ Formwork to comply with ACI 301, 318, and ACI 347.
_ Tolerance standards for level, plumb, and aligned construction shall be
   as per ACI 117.

PART 2 -- FORMWORK AND CONCRETE MATERIALS AND PRODUCTS

2.1 FORMWORK MATERIALS

*Note to specifier: If using formwork of materials other than wood, such as fiberglass, metal, fiberboard tubing, etc., refer to manufacturer's standards and/or ACI standards.

A. Formwork plywood shall be solid, with undamaged surfaces and edges.

_ Species: Douglas Fir.
_ Grade: Exterior, sanded both sides, PS 1(C).

*Note to specifier: Your choice of species depends on local custom and supply. Select grades and finishes as appropriate for visible and non-visible concrete finishes.

_ Formwork framing lumber:
_ Species: Douglas Fir.
_ Grade: Standard or better.

*Note to specifier: Your choice of species depends on local custom and supply. Douglas Fir, Pine, or Spruce are most common.

B. Formwork ties, anchors, braces, spacers, etc.

_ Formwork ties as required to prevent any dislocation in formwork.

*Note to specifier: Form ties may be removable, snap off, metal, fiber glass, etc. If ties are specified, select recommended or required manufacturer and refer to manufacturer's standards.

_ Shoring and braces as required to prevent any dislocation of formwork.

*Note to specifier: If braces are specified, select recommended or required materials or manufacturer and refer to ACI and/or manufacturer's standards.

_ Anchors and anchor slots   _ Hanger wires   _ Cans   _ Inserts
    _ Bucks _ Sleeves
    _ All as required to assure solid accessory support systems.
    _ All as required to avoid damage to concrete upon removal of
        formwork.

_ For concrete surfaces to remain visible in finished work:
_ Avoid damage to concrete surface from formwork.
_ Avoid damage to concrete surface from ties, braces, anchors and inserts.

*Note to specifier: If anchors, hangers, etc. are specified, select recommended or required materials or manufacturer and refer to ACI and/or manufacturer's standards.

_ Form spacers and ties removable so as not to leave metal close to
    the finish surface of concrete.
_ Types, materials and spacings as per applicable ACI standards.

C. Miscellaneous materials:

_ Flashing reglets: Galvanized steel as manufactured for this purpose with splines to align joints.

*Note to specifier: Reglets may be specified as galvanized steel or rigid PVC. If specified, select recommended or required materials or manufacturer and refer to ACI and/or manufacturer's standards.

_ Waterstops: Polyvinyl chloride.

*Note to specifier: Waterstops may be specified as polyvinyl chloride or rubber, polyvinyl chloride is most common. If specified, select maximum length, required width, profile, manufacturer(s), and refer to ACI and/or manufacturer's standards.

_ Construction joints: Tongue and groove extruded plastic as manufactured for this purpose.

*Note to specifier: Joints may be specified as extruded plastic or galvanized steel. If specified, select manufacturer and refer to manufacturer's standards.

_ Joint filler: Premolded asphaltic board as per ASTM D 1751.

_ Form-release agent: Clear mineral oil, non-staining, as manufactured for formwork.

*Note to specifier: If form release agent is specified, select manufacturer and refer to manufacturer's standards.

_ Vapor retarder for concrete slab on or below grade: 6 mil clear polyethylene.

2.2 REINFORCING MATERIALS

A. Reinforcing bars:

_ Deformed steel bars, Grade 60, Type S, to comply with ASTM A 615.
_ Plain finish bars may be used in spiral.

*Note to specifier: If bars are to be plain (non-deformed) as for spiral reinforcing, and galvanized finished, specify the applicable ASTM standards. Use galvanized or coated reinforcing if required to resist corrosion.

_ Fabrication to comply with CRSI--Reinforcing Bar Detailing.

*Note to specifier: Other standards of fabrication: ACI 318, ASTM A 184.

B. Welded wire:

_ Welded wire reinforcing:
_ Deformed to comply with ASTM A 497.
_ Plain to comply with ASTM A 185.

*Note to specifier: Specify flat or coiled wire reinforcing; finish plain, galvanized, or unfinished. If wire is to be plain (non-deformed) and galvanized finished, specify the applicable ASTM standards.

2.3 CONCRETE MATERIALS

A. Concrete ingredients:

_ Portland cement ASTM C 150 Normal-Type I.
_ Cement for concrete in contact with soil containing sulfates: Sulfate Resistant Type V.

*Note to specifier: Use Portland cement Type II for concrete in contact with soil containing moderate amounts of sulfates, a need to reduce the heat of hydration, and concrete exposed to sea water. Use Portland cement Type III if high early strength is required.

_ Aggregate, fine and course: ASTM C 33.

*Note to specifier: Aggregate for lightweight concrete shall conform to ASTM C 330.

_ Water as per ASTM C 94:
_ Clean, free of salt or any chemicals or contaminants that might injure the concrete.

*Note to specifier: Local drinking water will normally be acceptable unless there is a severe contamination by potentially damaging chemicals such as sulfates.

B. Admixtures and miscellaneous materials:

_ Air entraining admixture as per ASTM C 260 and manufacturer's instructions.

*Note to specifier: Select air entraining admixture as required to protect concrete from weather exposure. Product selection as limited by ASTM C 260.

_ Water reducing, retarding, accelerating admixtures as per:
  _ ASTM C 494.
  _ Manufacturer's instructions.

*Note to specifier: Select air entraining admixture as required for special conditions requiring retarded or accelerated curing of concrete. Product selection as limited by ASTM C 260.

_ Bonding agent: Polymer resin.

*Note to specifier: If bonding agent such as polymer resin emulsion or latex emulsion is specified, refer to manufacturer's standards.

_ Non-shrink grout:

    _ Non-metallic mineral aggregate, cement, water reducing materials as per ASTM C 494.
    _ As per manufacturer's instructions.

*Note to specifier: Non-shrink grout may be used to set dowels and anchor bolts.

2.4 CONCRETE MIXTURE

A. All mixing and tests to assure compliance with standards as per ACI 301.

_ Provide concrete ready-mixed in compliance with ASTM C 94.
_ On-site mixed concrete will conform to ASTM C 685.
_ Concrete strength will conform to ACI 301, 318, and applicable building code requirements.

_ Compressive strength of _____ psi in 7 day test.

_ Compressive strength of _____ psi in 28 day test.

_ Slump _____ to _____ inches.

_ Minimum water cement ratio:

*Note to specifier: Building code and engineer requirements must be consulted to establish the standards of concrete strength listed above.

_ Add air entraining admixture as required to protect concrete exposed to exterior weather.

_ Add admixture as per ACI 301 and 318 and manufacturer's instructions.

PART 3 -- CONSTRUCTION AND INSTALLATION

3.1 FORMWORK CONSTRUCTION

A. Install formwork:

_ At allowable distances from property lines.
_ At allowable distances from nearby adjacent construction.
_ As detailed for concrete member widths, depths, and heights.
_ Vertically plumb: ASTM 301 (4.3 -- Tolerances for Formed Surfaces).
_ Aligned at top of walls.
_ Square at corners.
_ Accessible for concrete placement equipment and workers.
_ Verify that scaffolds or other accessories are adequate and do not affect formwork.
_ Erect shoring and bracing in excess of what is required to firmly and fully support all loads.
_ Set cross bracings in formed openings to prevent deflection and bowing.
_ Camber slabs, girder, and beam framing to assure final level alignments:
_ Comply with ASTM 301 (4.3 -- Tolerances for Formed Surfaces).

_ Place, align, fasten, and protect formwork strips:

    _ Chamfers at external corners such as beams and columns.
    _ Nailers
    _ Rustication strips

_ Install forms so as to allow space and openings for flow and placement of concrete.
_ Construct formwork free of defects that would affect appearance of finish concrete surfaces.
_ Secure forms against movement during placement.
_ Secure forms against deflection.
_ Construct formwork to include required structural sections.

_ Construct walls, pilasters, column caps, etc. to have full bearing to support:

    _ Beams _ Joists _ Slabs

_ Set cross bracings in formed openings to prevent deflection and bowing.
_ Remove temporary spreaders.
_ Verify that scaffolds or other accessories are adequate and do not affect formwork.

B. Joints:

_ Install movement joints as detailed:

_ Secure fillers
_ Free movement isn't blocked
_ No reinforcement or fixed metal continues through movement joints

_ Install construction joint keys as detailed:

_ Locations _ Widths _ Keys _ Waterproofing

_ Construct form joints that are plumb and tight enough to prevent seepage.

C. Treating and cleaning formwork:

_ Clean formwork during construction to remove trash, scraps, and other foreign materials.

_ Inspect and clean formwork immediately prior to pour:
_ Clear of scraps and debris   _ Clear of organic matter
_ Clear of loose dirt
_ Clear of mud and water

_ Treat formwork:
_ Seal   _ Wet wood forms   _ Oil metal forms
_ Remove excess oil

_ Apply form-release materials according to manufacturer's requirements.
_ Apply before adding reinforcing and fittings.
_ Do not apply where such materials might damage applied finish materials.

_ Repair and recondition reused formwork so that strength, tightness, and surface match original.
_ Where concrete will be exposed, form boards must be free of stains or other contamination.

3.2 INSERTS, SLEEVES, OPENINGS

A. Coordinate with other work:

_ Provide openings, chases, sleeves, bolts, and inserts for other trades such as required for:
   _ Piping   _ Conduit   _ Ductwork   _ Foundation vents

_ Construct formwork that provides for all required:
   _ Depressed slab areas   _ Cutouts   _ Curbs   _ Inserts

_ Construct forms that provide for and are removable for:
   _ Doors   _ Windows   _ Vents   _ Access panel openings

_ Set floor components in coordination with finish floor elevations:
   _ Drains   _ Equipment anchors   _ Boxes   _ Cleanouts
   _ Flanges   _ Pipe sleeves

_ Leave consistently positioned access openings at the bottom of column forms on each floor.
_ Provide consistently aligned sleeves for plumbing and conduit at column forms on each floor.

B. Fittings, anchors, joints:

_ Install fittings, anchors, and other accessories, straight and plumb.
_ Install waterstops so that they are continuous and sealed, without breaks.
_ Waterstops or other accessories are not to be placed so as to disturb or dislocate reinforcing.
_ Separate slabs from walls and columns with 1/2 inch joint filler or bond breaker.
_ Install joint filler at joint lines.
_ Extend joint filler from bottom of slab to 1/2" of top of finished slab surface.

*Note to specifier: Joint filler as per ASTM D 1751.

C. Cleanouts, vents, and formwork inspection:

_ Provide cleanouts at ends and low points of forms.
_ Verify that the number and location of cleanouts are correct.
_ Vent potential pockets to prevent air entrapment.
_ Provide ports in high forms.

3.3 REINFORCING--INSTALLATION

A. Install reinforcing bars, wire, and related supports:

_ Provide support in excess of that needed to resist displacement during placement of concrete.
_ Install all reinforcing so it is never closer than 2" to sides or bottom of formwork.
_ Do not allow reinforcing to be in contact with dissimilar metals.
_ Create rebar spacings adequate to allow concrete flow and complete penetration and coverage.
_ Install stirrups and reinforcing at least 2" from side forms to prevent steel exposure at surface.
_ Where final surfaces will be exposed, use rust-free chairs and other formwork materials.

3.4 CONCRETE--PRE-PLACEMENT ADMINISTRATION AND COORDINATION

A. Coordination:

_ Confirm that all the latest design office construction documents are in use:
_ Architectural _ Civil _ Structural
_ Mechanical Shop Drawings as they may affect foundations with chases, sleeves, etc.
_ Electrical Shop Drawings as they may affect foundations with chases, sleeves, etc.

_ A preconstruction meeting will be held with all concerned parties.
_ Coordinate foundation layout, trenching, and formwork with:
_ Roof drainage _ Utility trenching and piping _ Foundation drainage

B. Permits and tests:

_ Obtain all required agency approvals.
_ Notify testing laboratory prior to pour.
_ Arrange testing at mix plant and at construction site.

3.5 PREPARATION FOR PLACEMENT OF CAST-IN-PLACE CONCRETE

A. Excavation:

_ Excavate footing trenches and prepare for pour:
_ Level _ Without soft spots _ Plumb
_ With firm and even side walls
_ Clear of debris _ Clear of loose dirt
_ Clear of organic matter
_ Clear of mud and water
_ Excavate all footing trenches below frost line.

B. Vapor retarder

_ Install vapor retarder under slabs on grade:

_ Joints lapped a minimum of 8 to 12 inches.
_ Any damage repaired and lapped and sealed over until watertight.
_ Seal entire vapor retarder watertight.

C. Coordination:

_ Coordinate installation of related work before concrete pour, and protect from damage:

_ Base plates _ Bolts
_ Sleeves for bollards
_ Utility boxes
_ Drain
_ Electrical conduit
_ Pipes and plumbing

_ Prepare previous concrete work for connection with new work:

_ Clean with wire brush.
_ Add bonding agent as per instructions of manufacturer.

_ Put required attachments, accessories, and inserts in place before pouring.
_ Install foundation formwork to allow for piping and conduit.

D. Inspection:

_ Complete and check all subgrade preparations prior to pour.

_ Verify that footing layout conforms to Drawings in:

_ Distances from property lines
_ All dimensions
_ Alignment _ Slopes

_ Make all necessary arrangements for continuous inspection of concrete during all phases of:

_ Batching _ Mixing _ Placing _ Finishing _ Curing

E. Joints:

_ Install joint filler at joint lines:

_ Separate slabs from vertical surfaces with 1/2 inch joint filler or bond breaker.
_ Extend joint filler from bottom of slab to within 1/2 inch of top surface of finished slab.
_ Install concrete without interruption between construction or expansion joints.
_ Pour floor slabs in a pattern of construction, expansion, and movement joints as per Drawings.
_ Screed floors and slabs level or slope as required by Drawings for drainage.

F. Protection:

_ Provide planks, or otherwise protect footing trenches to prevent
   dirt from falling in during pour.
_ Set up chutes or buggy walks to be moved during pouring to prevent
   overpours in one spot.
_ Keep all required tools on hand and in good working condition.
_ Areas to receive concrete shall be:
_ Free of debris or organic matter.   _ Wetted if dry.

3.6 MONITORING AND TESTING OF CAST-IN-PLACE CONCRETE

A. Testing materials:

_ Provide site sample test materials:

_ Cylinders   _ Slump cone   _ Measuring equipment

_ Provide concrete testing cylinders:

_ Mix and fill   _ Rodding   _ Strike off and tap
_ Curing under job conditions

B. Monitoring and record keeping:

_ Record dates and times of placement, interruptions, tests,
   completion, and finish work.

_ Maintain records for each pour:

_ Location   _ Extent   _ Date   _ Hour
_ Temperature   _ Weather

_ Verify concrete requirements before delivery and placement:

_ Tests   _ Mix design   _ Ingredients   _ Inspections

_ Slump must pass visual inspection.
_ Check test results at 3 or 7 days, and confirm at 28 days.
_ Install loose bolts and other insert items as detailed and as required for
   subsequent construction.

3.7 INSTALLATION

A. Work conditions:

_ Do not do work when rain or low temperatures might interfere
   with or harm the work.
_ Keep an adequate workforce on hand as required to handle the
   workload.

_ Check cold weather preparations:

_ Heating equipment and materials
_ Maintenance of heat in mix after placement

_ Check excavations and formwork during pouring to assure no dislocation or changes in:

_ Elevations _ Alignment _ Form adjustment
_ Formwork joint leaks
_ Formwork deflection or bowing
_ Supports, shoring, and braces

B. Mix and water:

_ Ready-mix concrete to conform to ACI 301 and 304:

_ Allowable time of transport
_ Mixing time
_ Certified as tested and/or inspected at the plant
_ Allow no unauthorized watering or overwatering.
_ Do not allow extended time to pass between addition of water and placement (as per in CSI 304).
_ Concrete mix must be rejected if there are signs of mix segregation.

_ Job-mixed concrete to conform to ACI and ASTM:

_ Keep cement in dry storage.
_ Sample sand and course aggregate for compliance.
_ Protect all materials from contamination.
_ Keep mix water clean and free of salts or other harmful chemicals.
_ Follow mixing time as specified.
_ Verify that mixing measurement tools are accurate.
_ Verify that test reports of cement and mixed concrete are satisfactory.

_ Follow a continuous concrete delivery schedule to allow uninterrupted placement.
_ Avoid any unplanned cold joints.
_ Have concrete delivered within specified time limits.
_ Carefully check and fill out all concrete delivery forms.

_ Do not allow mix trucks to stay beyond allowable waiting period before pouring concrete.

_ Typical waiting limits are:

_ Less than an hour on hot days
_ Less than half an hour after water has been added

C. Placement of concrete:

_ Avoid segregation of mix during pour:

_ Do not allow concrete to drop so far as to segregate or create voids.
_ Do not allow poured concrete to bounce across reinforcing bars.

_ Hold back pours to allow settlement of lower-level concrete.
_ Do not allow a holdback in pour to allow concrete to start initial set.
_ Provide grout at points of rebar interference.
_ Do not place concrete after initial set.
_ If necessary to change source of material, stop work until substituted materials can be approved.

D. Compaction and vibration:

_ Compaction or vibration is not to disturb formwork.
_ Keep vibrators and standby vibrators on hand and in good working condition.
_ Verify vibrator frequency and amplitude.

_ Vibrate concrete as it's placed:

_ With correct equipment
_ Without damaging formwork
_ Without striking reinforcing bars or other components
_ Only by trained and qualified vibrator handlers

E. Joints and lifts:

_ Install movement and relief joints as required by Drawings and as instructed by Architect:

_ At contact with other work.
_ At breaks in the construction sequence.

_ Prevent any damage to construction joints and movement joints.
_ Verify that layer pours are horizontal and within specified limits of lifts.

F. Formwork:

_ Keep formwork in place after pouring until concrete reaches required strength.
_ Adjust and retighten forms as necessary to fasten securely against concrete surfaces.

_ Keep all formwork bracing, shores, and supports in place after pouring:

_ Until ample time has passed for concrete to reach required strength.
_ Keep added loads on newly poured concrete well within safety limits.

3.8 REMOVING FORMWORK

A. Remove formwork as per CSI 301 and 318.

_ Remove wood formwork below grade, as well as above grade.
_ Concrete must reach sufficient strength to carry its own load, and imposed live and dead loads.
_ Cut off nails or other metal that extends below slabs in crawl spaces.

B. Damage prevention and repair:

_ Protect newly poured concrete surfaces from damage during and after stripping of forms.
_ Promptly remove form tie clamps before corrosion can begin.
_ Remove loose nails and other metals that might leave rust.
_ Grout any depressions in concrete smooth and level.
_ Do not remove bracing, shoring, or formwork until concrete has reached ample strength.

_ After form removal, promptly repair concrete surfaces that will remain visible, including:

_ Honeycombs _ Form marks _ Fins _ Holes
_ All other surface defects

C. Clean work surfaces, remove formwork, completely remove debris from the job site.

END OF SECTION

Notes:

DIVISION 3
CONCRETE

CURING AND FINISHING
03370

PART 1 -- GENERAL

1.1 WORK

A. Provide curing and finishing of concrete shown on the Drawings

_ ACI 305--Hot Weather Concreting.
_ ACI 306--Cold Weather Concreting.
_ ACI 308--Standard Practice for Curing Concrete.

PART 2 -- MATERIALS

2.1 CURING AND SEALING PRODUCTS

A. Curing and protection paper as manufactured by:

*Note to specifier: Name the manufacture, product trade name, and any related data required to clearly identify the intended product.

_ Products shall comply with ASTM C171.
_ Use nonstaining curing paper or paper with polyethylene film on floor slabs.

B. Liquid curing agents as manufactured by:

*Note to specifier: Avoid curing agents except where required because of unusual mix or weather conditions. Consult ACI standards noted under 1.2 A.

C. Slip-resistant abrasive-texture aggregate:

_ Aluminum oxide, grading 14/36.

D. Sealer as manufactured by:

*Note to specifier: Name the manufacture, product trade name, and any related data required to clearly identify the intended product.

PART 3 -- CONSTRUCTION

3.1 PREPARATION

A. All materials, equipment, and personnel shall be as required to perform the work shown and specified.

B. Verify that slabs will be properly sloped for required drainage.

3.2 CURING

A. Provide for curing of concrete as per ACI 308 for a minimum of seven days.

_ Start curing procedures promptly after pour, to protect concrete from premature drying.
_ Control curing methods, covers, and wetting, with special attention to weather conditions.
_ Carefully control wetting to match requirements imposed by weather conditions.
_ Use proper wet spray or moist curing methods as required and as appropriate to weather.
_ Apply waterproof paper or other cover with ample laps and seals, for complete curing protection.
_ Use special covers atop special concrete finishes or colored work as necessary to prevent stains.
_ Where formwork is exposed to sun, maintain moisture on formwork until removal.

B. During curing:

_ Protect concrete from heat or cold, to maintain temperature between 50 and 70F degrees.

_ Protect concrete from:
    _ Inclement weather or running water.
    _ Construction equipment.
    _ Shock.
    _ Movement or vibration.
    _ Load stress.

3.3 FINISHING

A. Match up finish work to adjacent or nearby surfaces at:

_ Joints _ Edges _ Corners

B. Joints:

_ Coordinate sawn joints, to keep all joints straight and continuous.
_ Keep joint lines uniform and free of damage.
_ Do not make any cuts in finished concrete that might affect structural integrity or strength.

C. Floating, troweling, and special finishes shall be as shown on the Drawings.

_ Finishes shall be as noted on the Drawings.
_ Do not begin floating until bleed water is gone.
_ Do not over-trowel.
_ Do not dust cement to expedite troweling start time.
_ Remove any marks left by finishing tools.

D. Finishes where shown on the Drawings:

_ Floated finish for surfaces to receive roofing.
_ Troweled finish for walking surfaces or those receiving floor covering or membrane.
_ Broom finish shall be light, medium or coarse, at the direction of the Architect.
_ Scratched finish for surfaces to receive cementitious material.
_ Non-slip finish for steps, landings, platforms, and ramps.

*Note to specifier: If specifying special formwork finishes, describe surface qualities desired in complete detail, and refer to appropriate ACI standards.

E. Surface tolerance:

_ After first floating, check plane of surface with 10' steel straightedge.
_ Finish work, measured with a 10' straightedge, must be:

*Note to specifier: Define tolerance of 1/8" in 10' in any direction for first class work, especially flooring; a true plane of 1/4" in 10' for intermediate grade work; and true plane of 1/4" in 2' as minimal utility grade work.

3.4. PROTECTION AND COMPLETION

A. Curing, protection, and sealing:

_ Provide ongoing wetting for curing as required by weather conditions.
_ Protect fresh slab work from foot or traffic damage.
_ Seal concrete surfaces where shown on the Drawings.
_ Provide one copy of sealer manufacturer's standard written warranty, if applicable.

3.5 REPAIR AND CLEANUP

A. Repair or replace work not in compliance with the Drawings or these Specifications.

_ Repairs shall be as directed by the Architect.

B. Clean work surfaces, and completely remove debris and excess materials from the site.

END OF SECTION

Notes:

_____

_____

_____

**DIVISION 3
CONCRETE**

SAMPLE CONCRETE CONSTRUCTION

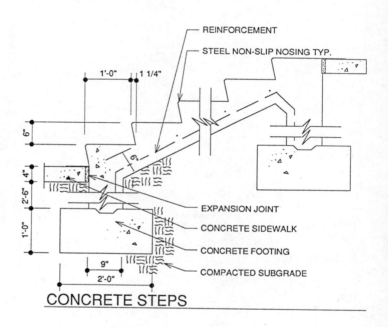

CONCRETE STEPS

DIVISION 4

MASONRY
04000

CONTENTS

04000	MASONRY	108
04210	BRICK MASONRY	111
04220	CONCRETE UNIT MASONRY	114
04230	REINFORCED UNIT MASONRY	115
	UNIT MASONRY -- PART 3 GENERIC TEXT	117
04400	STONE VENEER	123

*NOTE TO SPECIFIER: IN THIS DIVISION, ADAPT THE GENERIC PART 1, PART 2, AND PART 3 TEXTS TO THE SPECIFIC TEXTS PROVIDED FOR BRICK, CONCRETE UNIT MASONRY, AND REINFORCED UNIT MASONRY.

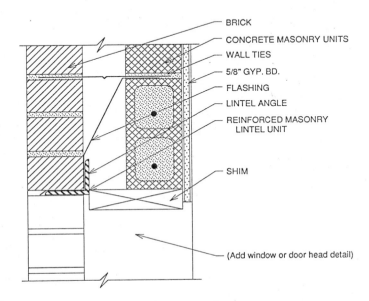

LINTEL @ 10" BRICK/BLOCK CAVITY WALL

DIVISION 4
MASONRY

MASONRY
04000

*Note to specifier: This is a generic PART 1 text that applies to all Unit Masonry Sections in this Division. Center the title and CSI number of the Specification Section in this title space.

PART 1 -- GENERAL

1.1 WORK

A. Provide everything required to complete unit masonry as shown on the Drawings and specified herein.

1.2 QUALITY STANDARDS

A. Provide experienced, well-trained workers competent to complete the work as specified.

B. Unless approved by the Architect, provide all related products and accessories from one manufacturer.

1.3 SUBMITTALS

A. Submit the following within _____ calendar days after receiving the Notice to Proceed.

*Note to specifier: Submittals are usually required within a specified number of calendar days after the Contractor is given the Notice to Proceed. 30 calendar days is a common requirement for medium-to-large-size projects. Your choice of time will depend on the size of the project and the Owner's need for an expedited schedule.

_ Submit list of materials to be provided for this work.
_ Submit manufacturer's data required to prove compliance with these specifications.
_ Submit manufacturer's installation instructions.
_ Submit shop drawings as required with complete details and assembly instructions.

*Note to specifier: Shop drawings may not be required if A/E drawings provide sufficient descriptive details for construction and installation.

_ Submit samples as required for approval by the Architect.

*Note to specifier: Samples may be requested with a different time limit than other submittals. Details of the samples requested -- sizes, finishes, etc. -- are usually specified.

1.4 MATERIALS HANDLING

A. Provide all materials required to complete the work as shown on drawings and specified herein.

_ Deliver, store, and transport materials to avoid damage to the product or to any other work.
_ Return any products or materials delivered in a damaged or unsatisfactory condition.
_ Materials and products delivered will be certified by the manufacturer to be as specified.

B. Store masonry off the ground, protected from dirt, ground moisture, contaminants, and weather.

1.5 PRECONSTRUCTION PROJECT PREPARATION

A. Examine and verify that job conditions are satisfactory for speedy and acceptable work.

_ Maintain and use all up-to-date construction documents on site.
_ Maintain and use up-to-date trade standards.
_ A preconstruction meeting will be held with all concerned parties.
_ Confirm there are no conflicts between this work and masons' local customs and building codes.
_ Do not begin to install masonry without approval of design firm.
_ Use agreed schedule for installation and for Architect's field observation.

PART 2   MATERIALS

2.1 MORTAR

A. Mortar as per ASTM C270.

_ Type S.
_ One part Portland cement.
_ One-half part lime.
_ Not more than four and one-half parts sand, measured damp and loose.
_ Compressive strength of 1800 psi at 28 days.

*Note to specifier: Type S is the general purpose mortar. Masonry used below grade or subject to high lateral or compressive loads or severe frost should use Type M, high-strength mortar with 2500 psi compressive strength. Minimal strength mortars of Type N and O may be allowable in the building code, but are not recommended. Mortar formula will vary depending on mortar types; see the Technical Notes of the Brick Institute of America for more information.

B. Mortar materials:

_ Portland cement: Type I or II.
_ Aggregate: Clean, sharp sand.

_ Lime:
    _ Quick lime: ASTM C5.
    _ Hydrated lime: Type S.
_ Water: Clean and potable.

C. Fireplace firebox: Use refractory mortar.

D. Coloring pigment: Provide pure ground mineral oxide, non-fading and alkali proof.

2.2 ACCESSORIES AND OTHER RELATED MATERIALS

A. Provide all accessories and materials as required for complete, proper installation.

B. Install reinforcing and anchorage as shown on the Drawings:

_ Reinforcing bars: Grade 40, unless otherwise shown on the Drawings.
_ Deformed bars for No. 3 and larger.
_ Single wythe joint reinforcement: Truss type.
_ Multiple wythe joint reinforcement: Truss type with moisture drip.
_ Joint reinforcement: Unprotected cold-drawn steel.
_ Strap anchors: Bent steel, 1/4" thick, galvanized.
_ Sheet metal wall ties: Corrugated galvanized steel.
_ Steel wire wall ties: Galvanized steel-formed wire.
_ Dovetail anchors: Bent strap, 1/4" thick galvanized steel.

C. Miscellaneous accessories and materials:

_ Building paper: No. 15 asphalt-saturated felt.
_ Flashing: non-corrosive sheet metal.

*NOTE TO SPECIFIER: CONTINUE WITH MATERIALS TEXT FROM SPECIFIC UNIT MASONRY SECTIONS THAT FOLLOW.

END OF SECTION

Notes:

_____

_____

_____

_____

_____

_____

_____

_____

DIVISION 4
MASONRY

BRICK MASONRY
04210

PART 1 -- GENERAL

*Note to specifier: Include generic introductory PART 1 -- GENERAL text from the beginning of Division 4.

PART 2 -- MATERIALS

*Note to specifier: Include generic PART 2 -- MATERIALS text from the beginning of Division 4. Then the MATERIALS text continues as follows.

2.3 BRICK MASONRY

A. Provide brick masonry as in the Drawings and specified herein.

B. Brick masonry will be manufactured by:

C. Brick masonry types and grades shall comply with ASTM C62 and C216.

_ Brick grade:

*Note to specifier: Brick grades are:

SW for severe weathering.
MW for moderate weathering.
NW for negligible weathering.

_ Brick type:

*Note to specifier: Brick types are:

FBX High mechanical quality, limited color range, minimal size variation per unit.
FBS Wide range of color, greater size variation per unit.
FBA Nonuniform in size, color and texture per unit.

D. Brick sizes, bond, and pattern as per drawings.

*Note to specifier: The most commonly used size, grade and type are as follows:

Modular size units, wire cut texture, grade MW, type FBS.

Standard Modular size: 3-3/4" x 2-1/4" x 8" with modular joint size of 3/8".

*Note to specifier:

Standard Modular size: 3-3/4" x 2-1/4" x 8" with modular joint size of 3/8".
Modular size: 3-3/4" x 2-1/4" x 7-5/8" with joint size of 3/8".

The Architect must always review samples of brick sizes, textures, and colors in the process of detailing and specifying.

_ Fire brick: ASTM C27 class, super-duty regular type.
_ Provide special units as manufactured for corners, jambs, sills, and lintels.

*Add the following text when using the generic PART 3 -- INSTALLATION portion of these Specifications for brick masonry:

_ Saturate bricks prior to bricklaying.

_ Determine that broken brick samples show the required moisture penetration:

    _ Core and perimeter are wet.
    _ Surface is dry.
    _ Brick isn't over-saturated.

PART 3 -- CONSTRUCTION AND INSTALLATION

*NOTE TO SPECIFIER: SEE THE GENERIC MASONRY PART 3 -- CONSTRUCTION AND INSTALLATION SECTION THAT FOLLOWS AND ADAPT IT TO THIS SECTION.

END OF SECTION

Notes:

DIVISION 4
MASONRY

SAMPLE BRICK MASONRY CONSTRUCTION

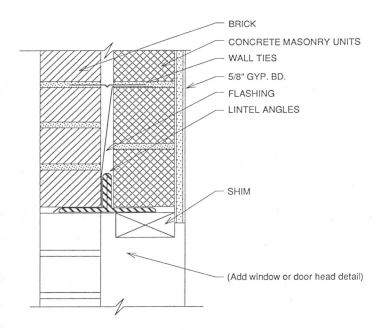

LINTEL @ 8" BRICK/BLOCK CAVITY WALL

DIVISION 4
MASONRY

CONCRETE UNIT MASONRY
04220

PART 1 -- GENERAL

*Note to specifier: Include generic introductory PART 1 text from the beginning of Division 4. Then the MATERIALS text continues as follows:

PART 2 -- MATERIALS

*Note to specifier: Include generic PART 2 -- MATERIALS text from the beginning of Division 4. Then the MATERIALS text continues as follows.

2.3 CONCRETE UNIT MASONRY

A. Provide concrete unit masonry where shown on the Drawings and specified herein.

B. Concrete unit masonry will be manufactured by:

*Note to specifier: Note here manufacturer's product identifications, trade names, catalog numbers, etc. Note the manufacturer's city and, if practical, the manufacturer's address, phone, and fax numbers.

C. Concrete unit masonry units shall comply with ASTM standards:

_ Concrete bricks, ASTM C55.
_ Hollow load-bearing units, ASTM C90.
_ Hollow non-load-bearing units, ASTM C129.
_ Solid load-bearing units, ASTM C145.

D. Grades and types:

_ Hollow, load-bearing units Grade N, Type I, Medium Weight for exterior.
_ Hollow, load-bearing units Grade S, Type II, Lightweight for interior.

E. Pattern and sizes shown on the Drawings.

_ Nominal size: 8" x 16" units as shown on drawings.

*Note to specifier: Standard American nominal size is 8" x 8" x 16"; actual size is 7-5/8" x 7-5/8" x 15-5/8" with 3/8" thick mortar joints.

*NOTE TO SPECIFIER: CONTINUE WITH "MATERIALS" TEXT ADAPTED FROM THE INTRODUCTORY GENERIC PART 1 AND PART 2 TEXT AT THE BEGINNING OF THIS DIVISION.

PART 3 -- CONSTRUCTION AND INSTALLATION

*NOTE TO SPECIFIER: SEE THE GENERIC MASONRY PART 3 -- CONSTRUCTION AND INSTALLATION SECTION AND ADAPT IT TO THIS SECTION.

END OF SECTION

DIVISION 4
MASONRY

REINFORCED UNIT MASONRY
04230

2.1 CONCRETE UNIT MASONRY

PART 1 -- GENERAL

*Note to specifier: Include generic introductory PART 1 text from the beginning of Division 4. Then the MATERIALS text continues as follows:

PART 2 -- MATERIALS

2.1 REINFORCED MASONRY

*Note to specifier: Reinforcing for reinforced brick masonry (RBM) should be detailed and specified by a qualified structural engineer.

"Low lift" reinforced brick masonry is constructed to a height of no more than 4' in each lift before grouting. This method is suited to smaller projects and hand application.

"High lift" reinforced brick masonry walls are grouted in one-story lifts. High lift is suited to engineered multistory projects that will be built with grout pumping equipment.

A. Preparation -- sample panel.

_ Construct sample wall panel of depth and height to show
    compliance with Drawings:
_ Coursing _ Joint sizes _ Bonding _ Uniform layering _ Patterning
_ Weep holes
_ Cleanouts _ Ties _ Reinforcing _ Parging _ Grouting

B. Grout:

_ Grout as per ASTM C476.
_ Grout compressive strength not less than 2000 psi at 28 days.

C. Tests and samples:

_ Schedule tests and inspections prior to construction.
_ Provide mill test certifications from suppliers.
_ Certify mortar is as specified regarding:
_ Tests _ Types _ Colors _ Mix and ingredients

_ Confirm positive results of laboratory tests:
_ Unit masonry _ Mortar mix ingredients
_ Grout ingredients _ Reinforcing

_ Use several approved masonry samples to establish acceptable range of variation.

_ Match masonry units to approved samples for:
_ Types _ Grades _ Sizes _ Shapes _ Colors
_ Textures _ Finishes _ Curing

_ Reject defective masonry units:
_ Cracks _ Kiln marks _ Size variation _ Bends _ Chips _ Patches
_ Spalling
_ Flaking _ Lime nodules

_ Conduct mortar tests to match site and work conditions.

D. Water

_ Protect water supply:
_ From freezing.
_ Maintain water clean and free of contaminants.

PART 3 -- CONSTRUCTION AND INSTALLATION

*NOTE TO SPECIFIER: SEE THE GENERIC MASONRY PART 3 -- CONSTRUCTION AND INSTALLATION SECTION THAT FOLLOWS AND ADAPT IT TO THIS SECTION.

END OF SECTION

Notes:

_____

_____

_____

_____

_____

_____

_____

_____

_____

DIVISION 4
MASONRY

*MASONRY -- PART 3
GENERIC TEXT

PART 3 -- CONSTRUCTION AND INSTALLATION

*Note to specifier: The following generic text can be applied for the PART 3 -- CONSTRUCTION AND INSTALLATION portion of the various preceding unit masonry sections.

3.1 WORK PREPARATION AND CONDITIONS

*Note to Specifier: You may include any or all of the following installation notes. Or to shorten the text you may exclude items you consider to be extraneous and rely primarily on line A., B., and C.

A. Complete this work in a timely fashion, without interfering with, or delaying the work of other trades.

B. Prepare all work according to applicable codes and regulations.

C. Prepare all work according to the standards and specifications of the Masonry Institute of America.

D. Moisture and climate control:

_ Do not work in freezing or inclement weather that might interfere with work quality.
_ Provide heat as necessary for proper, prior to installation and as necessary to prevent freezing.
_ Protect concrete masonry units from moisture, and keep them dry during installation.

*Note to specifier, use this reference for brick sections only:
Saturate bricks prior to bricklaying.

Determine that broken brick samples show the required moisture penetration:

Core and perimeter are wet.

Surface is dry.

Brick isn't over-saturated.

E. Work layout and preparation:

_ Prepare a work layout to establish and assure correct:

_ Coursing _ Patterns _ Elevation of base course
_ Opening sill and header heights.

_ Follow lateral structural support requirements as detailed.
_ Use mortar mix components according to trade standards, lab or site testing, and as detailed.
_ Obtain mortar compression test results as directed by the Architect.
_ Use wall thickness-to-height ratios as detailed.
_ Space and coordinate expansion/contraction joints to match building frame and thru-joints.

_ Prepare building foundations, lips, and sills that will support masonry to assure they are:

_ Clean _ Level _ Correct sizes

_ Check and correct as necessary any structural members that support masonry so they are:

_ Correctly located _ Plumb _ Aligned _ Braced

_ Install attachments that support masonry as per details.
_ Provide level shelf angles.
_ Provide bond breaks to allow differential movements between building frame and masonry wall.

_ Put in place, anchor, plumb, and level metal work that will be embedded in masonry:

_ Angles _ Bucks and frames _ Lintels

_ Provide flashing and openings for materials and fixtures to be installed.

_ Put appurtenances in place, anchoring, insulating, and protecting them from:

_ Impact damage _ Abrasion _ Chemical corrosion.

_ Appurtenances include but are not limited to:

_ Piping and conduit _ Ductwork _ Sleeves

_ Install bond-break felts, flashing, or other movement separators as detailed.

3.2 UNIT MASONRY INSTALLATION AND MORTAR APPLICATION

*Note to Specifier: You may include any or all of the following installation notes. Or to shorten the text you may exclude items you consider to be extraneous and rely on line A., B., and C.

A. Complete this work in a timely fashion, without interfering with, or delaying the work of other trades.

B. Complete all work according to applicable codes and regulations.

C. Complete all work according to the standards and specifications of the Masonry Institute of America.

D. Mortar joints:

_ Do mortar applications promptly.
_ Tooled or weathered joints:

*Note to specifier: Weathered, Vee or concave joints are recommended for maximum weather protection. Raked or struck joints are not recommended for walls exposed to severe weather. Deep stripped joints are not recommended, except for interior or otherwise fully weather-protected walls.

_ Construct mortar joint sizes as detailed.
_ Provide full head and bed joints.
_ Properly butter masonry unit edges.
_ Completely fill all joints: bed, cross, end, and head.
_ Do not tool joints prematurely before initial mortar set.
_ Tool joints without damaging mortar.
_ Promptly point holes, such as for line nails, as work proceeds.
_ Fully bed copings, blocks, and caps, and completely point joints.
_ Remove wedges as work progresses.
_ Repair defective units as work progresses.
_ Completely fill and level bed joints on lintels.
_ Lay brick courses in reference to a level line.
_ Align and plumb vertical joint lines in alternate courses
_ Keep wall face plumb and aligned story by story.

E. Install caulking, control joints, lintels, and flashing as detailed:

_ Keep caulking spaces at window and door frames uniform and of detailed size.
_ Keep spaces for expansion/contraction control joints uniform and of detailed size.
_ Recess window and door lintels from face of wall.
_ Tightly mortar parapet and fire wall flashing into walls.
_ Repoint counterflashing after roofers have turned it back over base flashing.

F. Control consistency and quality of materials and installation

_ Late-day work shall match quality of early-day work.
_ Late project and upper-story work shall match quality of early and lower-story or sample wall.

3.3 MASONRY ACCESSORIES AND REINFORCING -- INSTALLATION

A. Install metal ties for bonding as per details and referenced trade standards.
_ Assure compliance in types, sizes, spacing, depth of anchoring and corrosion resistance.

B. Install reinforcing as per details and referenced trade standards:

_ Stagger laps in bond beams.
_ Install door and window jamb hooks.
_ Provide lintel reinforcing for poured headers.
_ Provide pier and column ties.
_ Provide ties at top and bottom of vertical bars.
_ Provide all necessary bracing to guarantee safety during construction.
_ Provide extra reinforcement at corners of openings and intersections.
_ Do not deform reinforcing bars to force fit.

3.4 GROUTING

A. Preparation:

_ Complete required building department inspection of masonry before grouting.
_ Complete grout mix testing and certification before grouting.

_ Use accurate job grout mixing measuring units.
_ Measure by shovel load is not allowed.

_ Install inserts, anchor bolts, straps, dowels and bars as per detail drawings:
_ Foundation _ Headers _ Intersecting walls
_ Columns _ Spandrels
_ Floor lines _ Roof line

_ Clean starting beds just before grouting.
_ Clean all areas to be grouted and reinforced.
_ Close cleanouts before grouting.
_ Provide cleanouts at first course of each pour if using high lift.
_ Confirm high lift or low lift procedure.
_ Confirm detailed building code height limits for pours.
_ Provide ties, forms, and braces as necessary to prevent wall "blowout" during grouting.
_ Provide continuous supervision during high lift grouting.

B. Grouting operation:

_ Provide vibration or rodding as required for consolidation during pour.
_ Provide vibration or rodding approximately 15 minutes after pour to reconsolidate mix.
_ Fill bond beams completely.
_ Do not allow any grout leakage during pour.
_ Identify and check possible hidden points of grout leakage during pour.
_ Prevent dislocations of reinforcement during pour.
_ Keep grout level below tops of walls where keying is required.
_ Follow grout manufacturer's protective instructions when stopping work for an hour or more.

3.5 PARGING, AND WATERPROOFING

A. Parge and waterproof where shown in the Drawings:

_ Parge or otherwise treat backs of walls as required to receive backfill.
_ Do not perform backfilling prior to proper curing or additional wall support.
_ Use waterproofing manufacturer's recommended curing procedures.

3.6 WORK PROTECTION AND CLEANING

A. Clean all surfaces prior to work, during work shifts, and immediately upon completion:

_ Don't allow any mortar to enter expansion/contraction joints.
_ Don't allow mortar droppings on sills, copings, and projecting courses.
_ Clean wythes and other wall spaces so mortar does not provide water bridges across space.
_ Scrape mortar extrusions off inside wall.
_ Clean mortar droppings from brick anchors and straps, to avoid water bridges.
_ Immediately clean off spills on brick from poured-in-place concrete caps, copings, etc..
_ Clean beyond detection or replace any finish brick damaged by spilled concrete or mortar.
_ Clean out masonry refuse that accumulates within the building at each work shift.
_ Clean and wet down tops of uncompleted wall sections at start of each new work shift.
_ Protect tops of uncompleted wall sections at end of each full work shift.

3.7 REPAIR AND TOUCH-UP

A. After installation, inspect all work for improper installation or damage.

B. Repairs:
 _ Repair or replace any work damaged during installation.
 _ Repair work will be undetectable.

END OF SECTION

Notes:

_____
_____
_____
_____
_____
_____
_____
_____
_____

**DIVISION 4
MASONRY**

SAMPLE STONE VENEER CONSTRUCTION

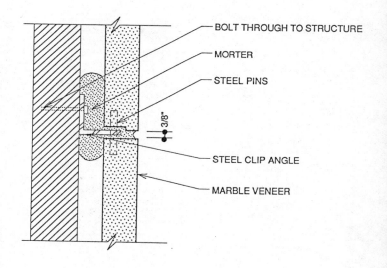

STONE VENEER @ HORIZ. JOINT

DIVISION 4
MASONRY

STONE VENEER
04400

*Note to specifier: Adapt the Part 1 GENERAL section from the preceding format data for use with stone.

_ Provide natural stone veneer as shown on the Drawings and specified herein.

2.1 MATERIALS

A. Mortar as per ASTM C270:

_ Portland cement: Type I or II.
_ Aggregate: Clean, sharp sand.
_ Lime:
  _ Quick lime: ASTM C5.
  _ Hydrated lime: Type S.

_ Water: Clean and potable.

B. Stone:

*Note to specifier:

_ Stone veneer 3 to 5 inches thick.
_ Stone pieces 1/2 sq. ft. to 1-1/2 sq. ft. and as approved by the Architect.

*Note to specifier: Select desired stone and grade from those listed below.

Limestone grades are of two colors, buff and grey, and four grades.
_ Select grade is fine-to-average grain with few flaws.
_ Standard grade includes fine-to moderate-size graining and some flaws.
_ Rustic grade has fine to very coarse grain.

Marble has four grade groups:
_ Group A, high-quality marble with uniform.
_ Group B, some faults and voids.
_ Group C, some flaws, voids, veins and lines of separation.
_ Group D, maximum of variation. Many of the most decorative colored marble is of Group D.

Stone grades in general:
Rubble: Irregular fragments with one good face.
Dimension: Cut in rectangular shapes:
> Large are called cut stone.
> Smaller blocks are called Ashlar.

Flagstone: Thin slabs suited to flooring and paving, either rectangular or irregular.

Granite may be as thin as 3/8".
Marble is usually 3/4" minimum.
Limestone, 2" to 3".

PART 3 -- CONSTRUCTION AND INSTALLATION

3.1 INSTALLATION

A. Lay stone as per manufacturer's or supplier's instructions:

_ Lay stone masonry with grain in horizontal position.
_ Nonstaining mortar is required.
_ Cover and protect work throughout construction.
_ Clean with mild soap and water and soft brush.
_ Never use acid cleaner.
_ Flashing of plastic or nonstaining metal.

3.2 REPAIR AND TOUCH-UP

A. After installation, inspect all work for improper installation or damage.

B. Repairs:

_ Repair or replace any work damaged during installation.
_ Repair work will be undetectable.

END OF SECTION

Notes:

_____
_____
_____
_____
_____
_____
_____
_____
_____

DIVISION 5

METALS
05000

CONTENTS

05300	METAL DECK	127
05100	STRUCTURAL STEEL FRAMING	130
05200	METAL JOISTS	136
05500	METAL FABRICATIONS AND MISCELLANEOUS METAL	139
05580	SHEET METAL FABRICATIONS	141
05700	ORNAMENTAL METAL	141
05800	EXPANSION CONTROL	141
05060	METAL FASTENING -- WELDING	142
05070	METAL FASTENING -- BOLTING	145

DIVISION 5
METALS

SAMPLE METAL STAIR CONSTRUCTION

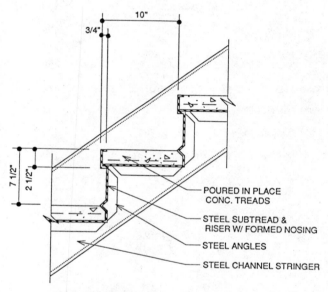

METAL STAIR: PAN TYPE

DIVISION 5
METALS

METAL DECK
05300

PART 1 -- GENERAL

1.1 WORK

A. Provide metal deck where shown on the Drawings and as specified herein.

1.2 QUALITY STANDARDS

A. All work must comply with applicable codes and regulations.

_ Deck used as part of fire-resistive assembly:
_ Must comply with applicable fire code.
_ Must comply with Underwriters' Laboratory requirements.

B. Provide experienced, well-trained workers to complete the work as specified.

C. Work must follow manufacturer's standards and instructions.

1.3 SUBMITTALS

A. Submit within ___ days of the Notice to Proceed.

*Note to specifier: Submittals are usually required within a specified number of calendar days after the Contractor is given the Notice to Proceed. 30 days is a common requirement for medium-to-large-size projects. Your choice of time will depend on the size of the project and the client need for an expedited schedule.

_ Submit list of materials to be provided for this work.
_ Submit manufacturer's data required to prove compliance with these Shop Drawings.
_ Submit manufacturer's installation instructions.
_ Submit Shop Drawings with complete details and assembly instructions.

1.4 MATERIALS HANDLING

A. Handle decking materials with care to avoid damage in:
   _ Transport _ Unloading and moving _ Stacking

_ Store decking materials as required to prevent damage:
_ Off the ground.
_ Protected from weather.
_ Protected from all sources of water.
_ Neatly stacked to prevent warping.
_ Secure from leaning or toppling.

PART 2 -- MATERIALS AND PRODUCTS

2.1 METAL DECKING

A. Deck as per AISC "Specification for Design of Light Gauge Cold-Formed Structural Members."

B. Material shall be galvanized as per ASTM A446, Grade A, and ASTM A525, designation G60.

C. Deck shall be as manufactured by:

*Note to specifier: Refer to specific manufacturer or choices of manufacturers of metal decking and follow manufacturer's recommended specifications. In most cases, your structural engineering consultant will provide recommendations and specifications for this work.

*Note to specifier: Add description of deck as per manufacturer's specifications.

_ Type:
_ Material:
_ Thickness:
_ Finish:
_ Method of attachment to joists:

*Note to specifier: Attachment will be either welded, bolted, or screwed.

2.2 ACCESSORIES AND OTHER RELATED MATERIALS

A. Provide accessories manufactured for use with the metal deck as specified.

_ Provide other accessories and materials as required for complete, proper installation.

PART 3 -- CONSTRUCTION AND INSTALLATION

3.1 PREPARATION AND PRECONSTRUCTION

A. Coordinate work that must be coordinated with metal decking.

_ Examine and verify that job conditions are satisfactory for speedy and acceptable work.

_ Coordinate work with related work of other trades:
_ HVAC _ Plumbing _ Electrical
_ Insulation _ Roofing

_ Provide closure pieces for:
_ Edges _ Tops of walls _ End points

_ Plan and provide supports and hangers for:
_ Suspended ceilings _ HVAC _ Piping _ Drains
_ Conduit _ Electrical fixtures _ Ceiling-hung equipment

_ Install well-secured, ample supports and hangers:
_ Tabs _ Inserts _ Anchors _ Wires

B. Steel-joist roofs shall:

_ Comply with fire-resistive requirements.
_ Comply with requirements as to combustible surfaces and finishes.
_ Use adequate bracing or bridging.
_ Be framed with sizes and types of joists as shown on Drawings and as specified.

3.2 INSTALLATION

A. Install in compliance with building code requirements and manufacturers' instructions.

_ Install decking so that it has continuous support and no unsupported edges.
_ Provide added support for decking around columns.
_ Provide added support for decking at openings.

_ Verify that joints and seams are:
_ Supported _ Straight _ Aligned

_ Install joint and seam connections as detailed:
_ Types _ Spacing _ Alignment to supports

_ Install built-in conduit passages:
_ Aligned _ Unobstructed _ Undamaged
_ Smooth at opening edges

_ Provide roof vents as shown on Drawings.
_ If exterior siding is part of decking contract, install all as per manufacturer's instructions.
_ Do not allow any heavy loads on decking.

3.3 REPAIR AND TOUCH-UP

A. After installation, inspect all decking to find damaged surfaces and repair damage.

_ Recoat any cuts in galvanized metal.
_ Touch up surfaces with galvanize repair paint approved by the Architect.
_ Repair or replace any other materials damaged during installation.

END OF SECTION

Notes:

_____

_____

_____

DIVISION 5
METALS

STRUCTURAL STEEL FRAMING
05100

*Note to specifier: Refer to specific manufacturer or choices of manufacturers of steel members and follow manufacturers' recommended specifications. In most cases, your structural engineering consultant will provide recommendations and specifications for this work.

PART 1 -- GENERAL

*Note to Specifier: In most cases, your structural engineering consultant will provide recommendations and specifications for this work.

1.1 WORK

A. Provide structural steel as shown on the Drawings and specified herein.

1.2 QUALITY STANDARDS

A. All work must comply with all applicable codes and regulations and:

_ AISC "Specifications for Design, Fabrication, and Erection of Structural Steel for Buildings"
_ AISC "Code of Standard Practice"

B. Provide experienced, well-trained workers to complete the work as specified.

1.3 SUBMITTALS

A. Submit within ___ days of the Notice to Proceed.

*Note to specifier: Submittals are usually required within a specified number of calendar days after the Contractor is given the Notice to Proceed. 30 days is a common requirement for medium-to-large-size projects. Your choice of time will depend on the size of the project and the client need for an expedited schedule.

_ Submit list of materials to be provided for this work.
_ Submit lab reports and any other data required to prove compliance with these Shop Drawings.
_ Submit Shop Drawings with complete details and assembly instructions.
_ Submit supplier's or manufacturer's specifications and installation instructions.

1.4 MATERIALS HANDLING

A. Deliver material to the job with components clearly marked as to location of installation.

_ Store materials protected to prevent damage and to maintain identification marks.

PART 2 -- MATERIALS AND PRODUCTS

2.1 STEEL

A. Steel and fastenings:

_ Steel plates, bars, and other shapes shall be as per ASTM A36.
_ Anchor bolts as per ASTM A307.
_ Nuts, bolts, washers as per ASTM A325.
_ Unfinished threaded fasteners as per ASTM A307, Grade A, regular.
_ Arc welding electrodes as per ASW code and ASTM A233, as
   required for intended use.

*Note to specifier: Rivets are rarely used in new steel construction. For specifications for riveted construction, consult your structural engineer.

2.2 PRIMER AND PAINT

A. Paint shall be:

Product   Manufacturer

*Note to specifier: Refer to specific manufacturer or choices of manufacturers of paint. Follow manufacturers' recommended specifications.

_ Primer: SSPC 15, type I, red oxide.

2.3. FABRICATION

A. Shop fabrication:

_ Fabricated structural steel as per AISC Shop Drawings and Shop
   Drawings.
_ Clearly mark materials for job site locations and sequence of assembly.
_ Fabricate according to most convenient delivery and erection sequence.
_ Complete all assembly and required welding before painting.
_ Finish surfaces must be clean and free of burrs and any other defects.

B. Connections:

_ Provide new, undamaged nuts, bolts, washers and other connectors as
   required.
_ Connectors as per ASTM A325.
_ Provide holes for connectors, anchors, etc. as shown on accepted Shop
   Drawings.
_ Provide openings for related work as shown on accepted Shop Drawings.

2.4 SHOP PAINTING

A. Paint structural steel surfaces except surfaces to be covered by concrete or mortar.

_ Apply two coats of paint to surfaces unreachable after erection.
_ Each coat of paint is to be a distinctly separate color.
_ Clean and prepare steel for painting as per standards of the Steel Structures Painting Council.
_ Apply primer and paint according to manufacturer's instructions and SSPC standards.
_ Primer and paint must be applied to achieve uniform thickness when dry.

PART 3 -- CONSTRUCTION

3.1 PREPARATION AND PRECONSTRUCTION

A. Job conditions and coordination:

_ Verify that job conditions are satisfactory for speedy and acceptable work.
_ A preconstruction meeting will be held with all concerned parties.
_ Review framing and coordination between trades during frame construction.

_ Maintain and use up-to-date construction documents:
_ Architectural   _ Structural framing
_ Steel fabrication Shop Drawings

_ Maintain and use up-to-date trade standards.
_ Provide ample adequate equipment for hauling, lifting, and securing steel framing members.

B. Coordination with other work:

_ Complete and coordinate rough plumbing with framing.

_ Cross-coordinate with framing plan:
_ Plumbing requirements.
_ HVAC requirements with framing plan.
_ Electrical requirements.
_ Foundation and pier construction.

_ Establish coordinated delivery and construction plan with steel supplier and subcontractors.

3.2 ERECTION

A. Comply with AISC Shop Drawings and standards and the instructions that follow.

B. Comply with applicable building code and standards referenced in building code.

C. Material tests and inspection:

*Note to specifier. Note locale and degree of required testing below.

_ Verify that fabrication is inspected and certified:
_ Shop _ Field _ Laboratory tested and certified

_ Obtain mill test reports with delivered steel, including heat numbers.
_ Maintain a log for identifying and tracking steel samples that are cut and sent for testing.

_ Delivered steel members shall include all types and quantities required for the project including:
_ Bearing plates _ Columns _ Pipe _ Lintels _ Girders _ Beams
_ Joists _ Cables _ Bracing _ Shoring _ Steel deck

_ Inspect steel deliveries, and reject:
_ Used steel mixed with new.
_ Substituted lighter, thinner materials.
_ Spliced or composited pieces not shown on Drawings.
_ Damaged materials.
_ Corroded material.
_ Material with nonconforming cuts or holes.

_ Provide steel framing for assembly with correct sizes and positions of:
_ Predrilled holes _ Cuts _ Connectors

*Note to specifier: You may use the following detailed requirements as explicit statements of exact standards required or, for shorter text, use the referenced standards A. and B. above.

_ Steel columns shall be:
_ New.
_ Free of damage or defects.
_ Straight.
_ Milled at ends if required for bearing.
_ Protected at milled ends.

_ Steel pipe and tubing shall be:
_ New.
_ Specified grades.
_ Free of damage or defects.
_ Straight.
_ Protected at ends.

_ Steel beams and girders shall be:
_ New.
_ Free of damage or defects.
_ Straight.
_ Rolled if specified.
_ Welded plate if specified or approved.
_ With specified camber.

_ Base and bearing plates shall be as detailed:
_ Sized at full required dimensions.
_ Fully bedded in grout.
_ Fit for anchor bolts.
_ All anchor bolts are fully connected.
_ Anchor bolts aren't bent during erection.
_ Concrete damage is minimal.

D. Bracing and temporary support:

_ Provide and install temporary supports:
_ With guys engineered to follow lines of force and not installed at off angles.
_ Securely braced.
_ With allowance for wind and storm forces in all seasons.
_ That do not move or shake when force is applied.

_ Securely reconnect members that have to be temporarily disconnected during erection.
_ Guy and brace all members securely to avoid a domino effect if any one member should fail.
_ Provide guys or braces at all sides of members that might move or fall during construction.

E. Erection:

_ Set column bolts with accurate templates.

_ Prepare concrete piers and footings where column base or bearing plates are installed:
_ Cleaned.
_ Grouted and rammed tightly.
_ Ram packed with dry pack mortar.
_ Shimmed with steel plate.

_ Erect vertical members:
_ Plumb within AISC tolerances.
_ Aligned to related members.
_ Undamaged in erection process.
_ With aligned and level column connectors for beams.
_ Erect horizontal members:
_ Place beams with cambers upward.
_ Match beam and column connection points.

_ Install bearing plates on masonry or concrete walls:
_ Exactly level.
_ Instrument checked.
_ Fully grouted.

_ Properly bed and connect end points of beams and girders bearing on masonry or concrete.

F. Connections:

_ Do not allow drift pins to damage steel members during erection process.
_ Temporarily fit all members in place, and before installing final connections.

_ Verify that members are:
_ Level   _ Plumb   _ Aligned

_ Use at least four bolts at each joint when setting heavier members temporarily in place.
_ Provide special connectors to other materials as per details and Shop Drawings.

G. Cuts and changes:

_ Make cuts sparingly and only as provided in Drawings for:
_ Plumbing _ Conduit _ Minor ductwork

_ Allow cuts to accommodate work of other trades only with written approval of the Architect.
_ Changes in structural framing are not allowed except through written approval of the Architect.

3.3 TESTING AND INSPECTION

A. Strength tests, lab tests, and field inspection:

_ The Owner's chosen testing lab will acquire specimens for any required tests.
_ Strength tests and inspections will be conducted as directed by the Architect.
_ Inspection shall be as directed by the Architect and by a qualified inspector chosen by the Owner.
_ Welding inspection will be by a qualified inspector from a testing lab chosen by the Owner.
_ Contractor must provide unrestricted access for testing lab representatives and inspectors.
_ Provide for additional testing to confirm correction of defective work.

B. If work is defective, the cost of additional testing and correction will be paid by the Contractor.

3.4 FIELD PAINTING

A. Apply specified paint according to manufacturer's instructions.

_ Apply specified paint to a dry film thickness of not less than 1.5 mils.

B. Do not paint:

_ Contact surfaces of high-strength bolts.
_ Steel that will be covered and protected by other materials, such as fireproofing or concrete.
_ Fabricated steel with adequate shop coatings.

END OF SECTION

Notes:

_____

_____

_____

DIVISION 5
METALS

METAL JOISTS
05200

*Note to specifier: Refer to specific manufacturer or choices of manufacturers of open web steel joists and follow manufacturers' recommended specifications. In most cases, your structural engineering consultant will provide recommendations and specifications for this work.

PART 1   GENERAL

1.1 WORK

A. Provide metal joists as shown on the Drawings and as specified herein.

1.2 QUALITY STANDARDS

A. All work must comply with applicable codes and regulations.

B. Provide experienced, well-trained workers to complete the work as specified.

C. Work must follow manufacturer's standards and instructions.

1.3 SUBMITTALS

A. Submit within ____ days of the Notice to Proceed.

*Note to specifier: Submittals are usually required within a specified number of calendar days after the Contractor is given the Notice to Proceed. 30 days is a common requirement for medium-to-large-size projects. Your choice of time will depend on the size of the project and the client need for an expedited schedule.

_ Submit list of materials to be provided for this work.
_ Submit manufacturer's data required to prove compliance with these Shop Drawings.
_ Submit manufacturer's installation instructions.
_ Submit Shop Drawings as required with details and assembly instructions.

1.4 MATERIALS HANDLING

A. Handle joists and trusses with care to avoid damage.

_ Reject and replace materials that have webs and chords with:
_ Bends   _ Dents   _ Cuts

_ Inspect and reject damaged joists and trusses after:
_ Transport   _ Unloading and moving   _ Stacking

_ Pick up and set joists in place according to manufacturer's instructions.

B. Storage.

_ Store joists and trusses so as to prevent damage:
_ Off the ground.
_ Protected from weather.
_ Protected from all sources of water.
_ Neatly stacked to prevent warping.
_ Secure from leaning or toppling.
_ Unload, store, and transport as per manufacturer's instructions and recommended equipment.
_ Store joists and trusses vertically on their chords.

PART 2 -- MATERIALS AND PRODUCTS

2.1 METAL JOISTS

*Note to specifier: Refer to specific manufacturer or choices of manufacturers of open web steel joists and follow manufacturers' recommended specifications. In most cases, your structural engineering consultant will provide recommendations and specifications for this work.

A. Joists will be as manufactured by:

*Note to specifier: Identify manufacturer or choices of manufacturers.

_ Types:
_ Materials
_ Depths:
_ Lengths:
_ Bridging:
_ Flanges:
_ End anchorage:
_ Concrete/masonry: Bearing plate.
_ Steel: Welded or bolted.
_ Joists shall be free of warps, twists, or bends.

*Note to specifier: Attachment will be either welded or bolted.

B. Joist accessories:

_ Provide accessories manufactured for use with the metal joists as specified.
_ Provide other accessories and materials as required for complete, proper installation.

PART 3 -- CONSTRUCTION AND INSTALLATION

3.1 PRECONSTRUCTION

A. Examine and verify that job conditions are satisfactory for speedy and acceptable work.

_ Clean joists and trusses before painting.

_ Coordinate joist and truss planning with all concerned disciplines and trades:
  _ HVAC _ Plumbing _ Electrical _ Roofing

3.2 FABRICATION AND INSTALLATION

A. Welding:

_ Shop and assembly welds as per Drawings:
_ Types _ Lengths _ Spacings

_ Verify welds:
_ Welders are certified.
_ Inspect visually.
_ Certified inspection.
_ Certified laboratory testing.

B. Installation:

_ Promptly attach bridging and anchors when placing joists and trusses.
_ Do not apply loads before attaching bridging and anchors.
_ Anchor at top and bottom chords, joist bridging that terminates at wall or beam lines.
_ Do not cut/drill chords and webs without approval of structural consultant.

_ Do not subject joists to heavy loads:
_ Materials storage.
_ Live loads from construction equipment.

_ Double check and retighten top-of-wall bolt connections.
_ Double check security of permanent truss bracing.

3.4 REPAIR AND TOUCH-UP

A. After installation, inspect all joists to find damaged members.

B. Repair or replace any damaged materials.

END OF SECTION

Notes:

_____

_____

_____

_____

_____

_____

_____

DIVISION 5
METALS

METAL FABRICATIONS AND MISCELLANEOUS METAL
05500

PART 1 -- GENERAL

1.1 WORK

A. Provide metal products as shown on the Drawings and as specified herein.

_ This work may include but is not limited to:

_ Handrails _ Metal fencing _ Metal stairs _ Ladders
_ Gratings _ Window gratings _ Floor plates _ Castings
_ Stair treads and nosings

B. Verify that job conditions are satisfactory for speedy and acceptable work.

1.2 QUALITY STANDARDS

A. All work must comply with applicable codes and regulations.

B. Provide experienced, well-trained workers to complete the work as specified.

C. Work must follow manufacturer's standards and instructions.

1.3 SUBMITTALS

A. Submit within ___ days of the Notice to Proceed.

*Note to specifier: Submittals are usually required within a specified number of calendar days after the Contractor is given the Notice to Proceed. 30 days is a common requirement for medium- to large-size projects. Your choice of time will depend on the size of the project and the client need for an expedited schedule.

_ Submit list of materials to be provided for this work.
_ Submit manufacturer's data required to prove compliance with these Shop Drawings.
_ Submit manufacturer's installation instructions.
_ Submit Shop Drawings as required with details and assembly instructions.

1.4 MATERIALS HANDLING

A. Handle materials with care to avoid damage in:

_ Transport _ Unloading and moving _ Stacking

_ Store decking materials as required to prevent damage:
_ Off the ground.
_ Protected from weather.
_ Protected from all sources of water.
_ Neatly stacked to prevent warping.
_ Secure from leaning or toppling.

PART 2 -- MATERIALS AND PRODUCTS

*Note to Specifier: This section is for miscellaneous metal products and fabrications as listed above under 1.1.A. Identify the type of metal products and refer to architectural drawings, Shop Drawings, and manufacturer's instructions.

PART 3 -- CONSTRUCTION AND INSTALLATION

*Note to Specifier: This section is for miscellaneous metal products and fabrications as listed above under 1.1.A. Identify the type of metal products, and refer to architectural drawings, Shop Drawings, and manufacturer's instructions.

END OF SECTION

Notes:

05580   SHEET METAL FABRICATIONS

*Note to Specifier: This section is for custom-made sheet metal enclosures. Use the generic three-part format, GENERAL, MATERIALS, and CONSTRUCTION / INSTALLATION, common to most other specifications sections. Identify the type and gauge of metal, and refer to architectural drawings and Shop Drawings.

05700   ORNAMENTAL METAL

*Note to Specifier: This section is for custom-designed ornamental metal such as:
   _ Ornamental Metal Stairs (05710)
   _ Prefab Spiral Stairs (05715)
   _ Handrails and Railings (05720)
   _ Metal Castings (05725)
   _ Ornamental Sheet Metal (05730)

*Use the generic three-part format GENERAL, MATERIALS, and CONSTRUCTION / INSTALLATION common to most other specifications sections. Identify the type and finish of metal and refer to architectural drawings for locations and dimensions, Shop Drawings for fabrication, and manufacturers instructions for assembly and installation.

05800   EXPANSION CONTROL

*Note to Specifier: This section may be included in Division 09000. Use the generic three-part format, GENERAL, MATERIALS, and CONSTRUCTION / INSTALLATION, common to most other specifications sections. Identify the type and gauge of sheet metal expansion joints, and refer to Drawings and manufacturer's instructions.

Notes:

_____

_____

_____

_____

_____

_____

_____

DIVISION 5
METALS

METAL FASTENING -- WELDING
05050

PART 1 -- GENERAL

1.1 WORK

A. Provide welded connections as shown on the Drawings and as specified herein.

*Note to specifier: Structural welding requires design, detailing, and specifying by a qualified structural engineer.

1.2 QUALITY STANDARDS

A. All work must comply with all applicable codes and regulations.

B. Provide experienced, well-trained workers to complete the work as specified.

C. Work must follow manufacturer's standards and instructions.

PART 2 -- MATERIALS

2.1 WELDING TOOLS AND MATERIALS

A. Materials as per standards of the American Welding Society:

_ Welding rod sizes and types.   _ Electrodes.   _ Cutting tool.

PART 3 -- CONSTRUCTION AND INSTALLATION

3.1 PREPARATION

A. Verify that job conditions are satisfactory for speedy and acceptable work.

B. Check potential sources of defective welds:

_ Chemicals.   _ Water.
_ Improperly prepared joint materials.
_ Improper connection of joint materials.
_ Sporadic starts and stops while welding a bead.
_ Lack of needed backup plates.
_ Proximity to thin metal members.

C. Before welding, inspect assembled materials, and repair or remove defective material:

_ Rust _ Scabs _ Seams _ Scale _ Plate laminations
_ Root openings   _ Proper edge preparation

3.2 SHOP WELDING

A. Shop welding shall conform to details and/or Shop Drawings and standards of the AWS.

_ Document the testing and certification of shop welding.
_ Document that welders are certified.

3.3 WORK CONDITIONS

A. Establish working conditions that conform to industry standards:

_ Temperature is within working range.
_ Do not allow water on metal surfaces.
_ Strictly enforce worker safety regulations.
_ Strictly enforce fire safety regulations.

B. Welders:

_ Allow welders to do only work covered by their certification.
_ Keep a field log of welders' names, work marks, and rejection rates.
_ Maintain a complete welding record that includes:
_ Pieces welded.
_ Weld locations.
_ Welder identification.
_ Location and extent of defects.
_ Repair dates.
_ Repair types.
_ Retests.
_ Sampling rate.

_ Provide testing laboratory field inspection and certification.

3.4 CONSTRUCTION

A. Application

_ Make sure materials to be welded are:

_ Securely clamped together.
_ Tack-welded if required.

_ Certify that welds are completed in out-of-view areas.

_ Before welding, inspect assembled materials, and repair or remove defective material:

_ Rust _ Scabs _ Seams _ Scale _ Plate laminations
_ Root openings _ Proper edge preparation

_ Supervise welding operations to confirm:

_ Correct preheat temperatures.
_ Single-pass welds are correct.

B. Inspection and repairs:

_ Continuously observe and inspect multi-pass welds.

_ Provide finish welds that conform to details and Shop Drawings in:
_ Types _ Sizes _ Lengths _ Locations _ Spacings
_ Contour _ Lack of defects

_ Provide nondestructive testing (NDT) whenever required by the Architect:
_ Radiography _ Ultrasonic _ Magnetic particle
_ Penetrant test

_ Redo all defective welds that show unacceptable:
_ Cracks _ Undercuts _ Pockmarks _ Holes _ Overlaps
_ Delaminations _ Tearing _ Warps _ Incomplete penetration
_ Incomplete fusion _ Burn-damaged structural members

_ Correct all work marked as nonconforming.

_ Document repairs.

_ Correct all repaired welds marked as unsatisfactory.

_ Reopen any work that has been closed in prior to inspection by the Architect.

END OF SECTION

Notes:

_____
_____
_____
_____
_____
_____
_____
_____
_____
_____
_____

DIVISION 5
METALS

BOLTING
05050

PART 1 -- GENERAL

1.1 WORK

A. Provide bolted connections as shown on the Drawings and as specified herein.

*Note to specifier: Structural bolting requires design, detailing, and specifying by a qualified structural engineering consultant.

1.2 QUALITY STANDARDS

A. All work must comply with all applicable codes and regulations.

B. Provide experienced, well-trained workers to complete the work as specified.

C. Work must follow manufacturer's standards and instructions.

PART 2 -- MATERIALS

2.1 BOLTS

*Note to specifier: Types and sizes of bolts must be as per Drawings and Specifications by a qualified structural engineer.

PART 3 -- CONSTRUCTION AND INSTALLATION

3.1 BOLTED FASTENING

A. Make bolt holes as detailed:
_ Sizes _ Locations _ Alignment _ Spacing patterns
_ Tolerances
_ Spacing of edge and end distances

_ Align and temporarily secure matched bolt holes at connections before bolting.
_ Do not burn or widen holes to force an alignment.
_ Ream or otherwise correct misfit holes only in a manner approved by structural consultant.

B. Corrections and quality control:

_ Replace bolt shanks that are bent during installation.
_ Replace bolt threads damaged during installation.
_ Fit bolt heads and nuts evenly against connected metal.
_ Draw bolts up tight.
_ Retighten all bolts before work is closed in.

_ Verify that there are no missing nuts and bolts in out-of-view areas such as:
_ Tops of masonry walls
_ Top plates of columns
_ Back sides of base plates
_ Out-of-reach sections of beam/joist connections

_ Complete field painting of metal before closing in.
_ Do not allow damage to bolt threads.

C. Inspection and repairs:

_ Reopen any work that has been closed in prior to inspection by the Architect.
_ Repair defective work as directed by the Architect.

END OF SECTION

Notes:

DIVISION 5
METALS

SAMPLE METAL FABRICATION CONSTRUCTION

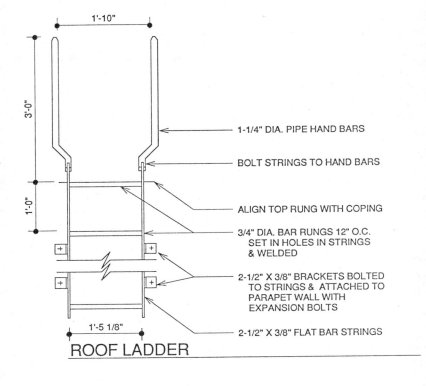

DIVISION 6

WOOD AND PLASTIC
06000

CONTENTS

06010	LUMBER AND ROUGH CARPENTRY	150
06200	FINISH CARPENTRY AND MILLWORK	163

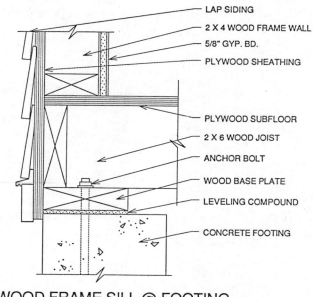

WOOD FRAME SILL @ FOOTING

DIVISION 6
WOOD AND PLASTIC

SAMPLE TIMBER CONSTRUCTION

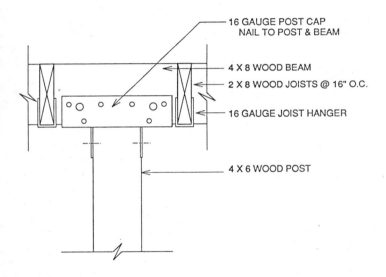

WOOD POST @ BEAM

DIVISION 6
WOOD AND PLASTIC

LUMBER AND ROUGH CARPENTRY
06010

PART 1 -- GENERAL

*Note to specifier: This is a generic PART 1 text that applies to all Sections in this Division.

1.1 WORK

A. Provide wood framing and finish carpentry as shown on the Drawings and as specified herein.

B. Where additional instruction is required, work shall be as directed by the Architect.

C. Work includes all lumber, connectors, and related hardware and materials.

1.2 SUBMITTALS

A. Provide all tests, certificates, and affidavits necessary to verify materials are as specified in:

_ Species and grades   _ Water content   _ Wood preservative

1.3 QUALITY STANDARDS AND TOLERANCES

A. Provide a work force that is:

_ Sufficient in number for the work load and time schedule.
_ Capable of handling any special heavy-duty or high-lift operations.
_ Skilled, trained, experienced, and competent to do the work as specified.

B. Unless otherwise directed by the Architect, all work shall be as per building code and:

_ Manual for Wood Frame Construction, American Forest and Paper Association (NFPA).
_ National Design Specifications for Wood Construction of the NFPA.
_ Plywood Specifications and Grade guide of the American Plywood Association.

*Note to specifier: If using Southern Pine, redwood, or other lumber, make reference to the appropriate trade association standards and grading rules.

C. Tolerances:

_ Vertical framing shall be plumb within 1/4" per 10 linear feet.
_ Horizontal framing shall be level within 1/4" per 10 linear feet.

D. Moisture contents and tests:

_ Moisture content of framing lumber shall be 19% or less by weight.
_ Tests will be conducted on all newly shipped lumber to confirm moisture content.
_ Kiln-dried or other lumber requiring lower moisture content shall be as specified.

E. Grading:

_ Follow applicable lumber grading agency standards in accepting or rejecting delivered lumber.

*Note to specifier: Refer to the standards of the American Forest and Paper Association or, if using Southern Pine, redwood, or other lumber, make reference to the appropriate trade association standards and grading rules.

_ Reject special, required lumber that is not marked and certified as:
_ Preservative-treated   _ Kiln-dried

_ Do not accept or use wood panels that deviate from grade standards or have excessive:

_ Marred surfaces   _ Cracks   _ Defective patches
_ Loose knots   _ Split edges
_ Delaminations

1.4 MATERIALS HANDLING AND STORAGE

A. Delivery and inspection:

_ Reject any delivered framing lumber that is not grade-stamped.
_ Verify that delivered lumber is grade certified by a bona fide grading agency.
_ Identify framing lumber by grade, and store each grade separately.

_ Do not accept or use lumber that deviates from grade standards or has excessive:

_ Moisture content   _ Loose knots   _ Decay streaks
_ Rot   _ Insect damage   _ Splits   _ Pitch pockets
_ Wanes   _ Crooks   _ Warps   _ Twists   _ Bends

_ Reject special, required lumber that is not marked and certified as:

_ Preservative-treated   _ Kiln-dried

_ Do not accept or use wood panels that deviate from grade standards or have excessive:

_ Marred surfaces   _ Cracks   _ Defective patches
_ Loose knots   _ Split edges   _ Delaminations

_ Remove unstamped or defective lumber from the job site.

B. Handling:

- Handle lumber to avoid damage during transport, unloading, and moving on the job site.
- Reject and replace lumber that is damaged and made unusable in off-loading during delivery.
- Handle chemically treated lumber and panels strictly according to manufacturers' instructions.

C. Storage:

- Store framing lumber and wood panels to prevent damage and moisture absorption:

- Well supported off the ground.
- Protected from weather.
- Away from traffic.
- Protected from all sources of water.
- Neatly stacked to prevent warping.
- Stacked with cross pieces for ventilation.
- Shored and with level support to prevent leaning or toppling.

- Store metal connectors that are subject to damage:

- In safe locations away from traffic or other sources of damage.
- In weathertight wrapping.

- Store chemically treated lumber and wood panels outdoors until installation.
- Keep chemically treated lumber and wood panels well ventilated if stored indoors.

PART 2 -- MATERIALS

2.1 FASTENERS, CONNECTORS, AND SUPPORTS

A. Hot-dip galvanized steel for exterior, high humidity, and treated wood locations.

B. Nails:

- Common wire or spike nails as shown on nailing schedule.
- Follow all nail size requirements and nail spacings required by the governing building code.
- Use hot-dip galvanized steel nails at exterior work, areas of high humidity or at treated wood.
- Plain finish materials may be used at interior and dry locations.
- Electro-galvanized nails shall not be used on exterior surfaces.
- Electro-galvanized nails shall not be used where corrosive staining might mar wood surfaces.
- Nails into redwood or cedar shall be of stainless steel.
- Aluminum nails for exterior work may be used at Contractor's discretion.

C. Power-driven nailing: Comply with standards of the International Staple, Nail and Tool Association.

D. Bolts:

_ Machine bolts to comply with ASTM A307.
_ Lag bolts to comply with Federal Spec FF-N-1.
_ Drill holes 1/16" larger than bolt diameters.
_ Drill straight through from only one side.
_ Use washers under all nuts.
_ Do not bear bolt heads on wood; use washers.

E. Hangers, connectors, and crossbridging.

*Note to specifier: A common product specification is: "Teco, Simpson, or equal as approved by the Architect."

_ Joist Hangers
_ Metal framing connectors.
_ Metal crossbridging.
_ Galvanized steel, sized to suit framing.

F. Anchors to adjacent construction:

_ Hollow masonry: Use toggle bolt.
_ Solid masonry or concrete: Use expansion shield and lag bolt.
_ Steel: Use bolt or ballistic fastener.

2.2 LUMBER

*Note to specifier: Select the lumber species and grade according to design needs or prevailing local custom. Light framing grades will be Construction, Standard, or Stud.

A. S4S, S-Dry unless otherwise indicated, grade marked complying with the following:

_ Girder Framing:
_ Species:
_ Grade:

_ Joist framing:
_ Species:
_ Grade:

_ Studs (2 to 4 inches thick or wide, 10 feet in length or shorter):
_ Grade: "Stud" or No. 3 Structural Light Framing.

_ Rafter framing:
_ Species:
_ Grade:

_ Non-structural light framing:
_ Species:
_ Grade: Standard or better.
_ No Utility grade.

_ Sill boards: Pressure treated or redwood sill grade.
_ Structural light framing: No. 2 or better.

_ Lumber for miscellaneous applications shall be Standard grade unless noted otherwise for:
_ Bucks _ Nailers _ Grounds _ Stripping _ Blocking
_ Furring _ (and similar)

2.3 SHEATHING AND UNDERLAYMENT -- MATERIALS

A. Sheathing and underlayment:

_ Plywood sheathing: Use APA rated, PS-1 or APA PRP-108.
_ Particleboard: Exterior Type 2-M.
_ Hardboard: ANSI/AHA A135.6.
_ Oriented Strand Board (OSB).
_ Subflooring: APA rated plywood sheathing, Exterior Grade.
_ Roof sheathing: APA rated plywood, Exterior Grade.
_ Underlayment: APA rated underlayment, Exterior; or:
_ Particleboard, Oriented Strand Board, or waferboard with waterproof resin binder.

B. Related construction and materials:

_ Sill gasket atop foundation wall: Glass fiber strip with width equal to plate.
_ Sill flashing: Galvanized steel or aluminum.
_ Subfloor glue: APA AFG-01, solvent base, waterproof.
_ Building paper: No. 15 asphalt felt (or spun-bonded polyethylene).
_ Vapor barrier: 6 mil polyethylene.
_ Termite shield: Galvanized sheet steel or aluminum.

2.4 WOOD TREATMENT

A. Wood preservative:

_ Provide wood preservative as follows:

Type	Color	Location	Manufacturer:

_ Pressure treatment: AWPA Treatment C.

_ Waterborne preservative with 0.25 percent retainage, rated for specific uses noted on Drawings.

PART 3 -- INSTALLATION

3.1 WOOD FRAMING -- PREPARATION AND PRECONSTRUCTION

A. Examine and verify that job conditions are satisfactory for speedy and acceptable work.

B. Coordination:

_ Maintain complete files of up-to-date design documents at the job site:
_ Architectural _ Structural framing
_ Wood fabrication Shop Drawings
_ Consulting engineer drawings that may affect framing such as plumbing and HVAC.

_ Maintain and refer to the latest trade standards.
_ Coordinate and complete rough plumbing before starting framing.
_ Cross-coordinate plumbing requirements with framing plan.
_ Cross-coordinate HVAC requirements with framing plan.
_ Cross-coordinate electrical requirements with framing plan.

_ Identify actual dimensions of all required rough openings in framing:
_ Doors _ Windows _ Other framed openings

_ The Architect will schedule a preconstruction coordination meeting with all concerned parties.

C. Operations:

_ Provide framing and shoring plan and schedule.
_ Provide lifts or cranes to assist high-level framing.

_ Verify that materials are stored so as to not overload or interfere with construction in terms of:
_ Quantities and weights _ Locations _ Traffic

3.2 ROUGH CARPENTRY, WOOD FRAMING -- AT GRADE AND FOUNDATIONS

A. Preservation, termite treatment, and ventilation:

_ Apply termite prevention where untreated wood will be within 8" of finish grade of soil.

_ Use foundation grade or preservative-treated lumber:
_ Near soil.
_ In contact with concrete.
_ In contact with masonry.
_ In spaces subject to concentrated moisture.
_ For all mudsills and precast concrete pier caps.
_ Do not use untreated wood wedges or shims in any location subject to moisture and decay.
_ Provide ventilation space for girders that will be set in foundation wall pockets.

B. Installation:

_ Install foundation wall sills and cripples as per Drawings and according to building code:
_ Plumb framing.
_ Square corners.
_ Top-of-plate elevations correct and consistent.
_ Level plates.
_ Plates and cripples aligned vertically and horizontally.

_ Position foundation anchor bolts so that none are underneath any studs.

_ Shims for mudsills shall be:
_ Wood shims equal to foundation grade or preservative- treated lumber.
_ Use steel shims for multistory construction.

C. Completed mudsills shall be:

_ Straight with a side variation tolerance of 1/4" per 10 linear feet.
_ Level within 1/4" per 10 linear feet.

3.3 ROUGH CARPENTRY, FRAMING MEMBERS -- FLOOR JOISTS

A. Install floor framing members as per framing plan, details, and as required by the building code:

_ Grades _ Sizes and spacings _ Bracing
_ Minimal notching or drilling
_ Install floor joists:
_ Set with crowns set upwards.
_ Set with full bearing on plates.

_ Install double floor joists under parallel partitions.
_ Install floor framing for concentrated floor loads including:
_ Close joist spacing _ Double or triple joists
_ Girder supports _ Blocking

_ Install double floor joists and blocking where framed around floor openings.

B. Install joist hangers as per Drawings, manufacturer's instructions, and building code requirements.

_ Set straight. _ Aligned.
_ Completely secured at all connection points.
_ Secured with correct size and type fastenings.

3.4 ROUGH CARPENTRY, WOOD FRAMING -- EXTERIOR AND INTERIOR WALLS

A. Install stud framing as per framing Drawings and building code requirements:

_ Plumb _ Square _ Aligned _ Substantially braced.
_ Secured with correct sizes and types of fastenings.

_ Install fire stops so as to provide complete, snug blocking between studs.

_ Install special framing as required for:
_ Double walls at chases.
_ Separate plates and framing at party walls.
_ Separate plates and staggered studs at soundproof walls.

_ Position studs at corners to provide ample nailing backing for interior and exterior panels.
_ Provide blocking and double top plate headers for wall openings.
_ Lap top plates and set butt joints so they don't occur over openings.

_ Install top plates to provide uninterrupted, ample nailing backing for
   interior and exterior panels.
_ Install headers and lintels as per details and building code with:
_ Ample bearing.
_ Secure connection to supports.
_ Provide complete and secure temporary bracing:
_ Nailing and stop plates at floors and slabs.
_ Double-sided prop bracing at walls.
_ Diagonal horizontal cross bracing at plates of intersecting walls.
_ Braced walls won't move, waver, or shake when force is applied to them.

B. Framing for related work:

_ Install furring as per Drawings and manufacturer's instructions.
_ Prepare stud framing for soundproofing as detailed.
_ Prepare stud framing for waterproof finishes as detailed.
_ Construct stud framing and blocking to support wall-mounted fixtures,
   cabinets, and equipment.

3.5 ROUGH CARPENTRY, WOOD FRAMING --
     CEILING AND ROOF

A. Install ceiling and roof framing members as per framing plans, details,
   and building code requirements:

_ Install ample bracing   _ Minimal notching or drilling

_ Install ceiling and roof joists:
_ Set with crowns upward.
_ Set with ample bearing on plates.
_ Securely anchored to plates.
_ Provided with more tie downs than building code requires if subject to
   severe winds.

_ Install rafters and sloped roof joists:
_ Coordinated with roof drain design.
_ Sloped for positive roof drainage.

_ Make angled rafter cuts that are:
_ Tightly fitted.
_ Securely anchored.

B. Framing for related work:

_ Install ceiling soffits and furring as per Drawings and Architect's
   instructions:

_ Prepare framing for soundproofing as detailed.
_ Install double roof joists and blocking where framed around roof
   openings.

3.6 SUBFLOOR SHEATHING

A. Installation:

_ Install plywood subflooring as per framing drawings and building code requirements:

_ Staggered pattern.
_ Nailing pattern.
_ Blocking with 100% support at all edges and support as required at intermediate spans.

_ Stagger subflooring butt joints.
_ Install subflooring panels so that edges have full bearing on framing members.

B. Fastening:

_ Glue and secure subflooring to floor joists with screw-type nails.
_ Subfloor-to-joist connections must be sufficient to totally prevent any squeaking of flooring.
_ Prepare framing for soundproofing as detailed.
_ Prepare framing for floor-mounted fixtures and equipment.

C. Completed subflooring shall be:

_ Level within 1/4" per 10 linear feet.
_ Free of depressions or humps.
_ Patched to repair holes, splits, or construction damage.

3.7 SHEATHING, SIDING, AND FINISH-UP WORK

A. Installation:

_ Stagger wall sheathing butt joints.
_ Install wall sheathing panels so that edges have full bearing on framing.
_ Install plywood shear wall construction as per the Drawings and as required by building code.

_ Comply with building code requirements for:
_ Thicknesses of plywood   _ Nail types and sizes
_ Nailing pattern

_ Install siding so that joints:

_ Are square.
_ Are staggered in alternate pieces if so designed.

_ Include 1/8" expansion joints between sheathing panels.

B. Finishes:

_ Prepare plywood surfaces for paint or stain according to manufacturer's instructions on:

_ Preservatives _ Patching _ Sanding
_ Cleaning _ Priming

3.8 WOOD FRAMING -- COORDINATION

A. Coordination with other work -- utilities, fixtures, equipment, finishes:

_ Coordinate electrical stub-ups with the framing plan.
_ Align floor-mounted electric outlet boxes with finish wall lines.

_ Coordinate girders, floor joists, and stud walls with plumbing:

_ Supply lines _ Floor drains _ Thru-building roof drains

_ Coordinate girders and floor joists with HVAC ducts and vents.
_ Do not allow HVAC ducts in wall framing to protrude beyond face of framing.

_ Recess floor joists to allow for:

_ Underlayment _ Tile flooring _ Poured topping
_ Recessed mats _ Recessed grilles
_ Changes in floor surfaces

_ Supply and coordinate in-wall fixture and equipment supports:

_ Anchors _ Brackets _ Grounds _ Chairs _ Frames

_ Provide in-wall blocking, anchors, brackets, grounds, and other supports for wall-supported:

_ Plumbing fixtures _ Electrical fixtures _ HVAC equipment
_ Kitchen or shop equipment _ Bathroom accessories
_ Handrails _ Guards/protective rails _ Shelves
_ Storage units _ Fire hose cabinets

_ Provide curbs, anchors, brackets, pitch pockets, and other supports, to support roof-mounted:

_ Platforms _ Piping _ Plumbing vents _ Light fixtures _ Roof vents
_ HVAC equipment _ Room enclosures _ Railings
_ Communications equipment _ Guy wires

_ Install plaster grounds as detailed and as per trade association standards.

B. Movement joints and clearances:

_ Provide joints and connectors for non-wood construction to allow for movement such as:

_ Lumber shrinkage _ Concrete shrinkage
_ Masonry expansion and contraction
_ Overall building thermal expansion and contraction.

_ Provide clearances between framing and other construction that may be subject to:

_ Differential movement _ Noise transfer

_ Provide clearances between framing and other construction subject to fire hazard such as:

_ Chimneys and flues
_ Thru-building expansion joint
_ Elevator cores

C. Waterproofing, water barriers, and vapor barriers:

_ Set and prepare framing as required for tile or other waterproof wall finishes.
_ Provide waterproofing sealing as detailed.
_ Combine water barriers with framing as shown on Drawings.
_ Water barriers, vapor barriers, and flashing must be undamaged.
_ Install water barriers, vapor barriers, and flashing as per manufacturers' instructions.
_ Combine vapor barriers as shown on Drawings.
_ All flashing must have unobstructed drainage to exterior.
_ Prepare framing for waterproof finishes where waterproofing required.

D. Insulation and sound barriers:

_ Combine thermal insulation with framing as shown on Drawings.
_ Combine soundproofing with framing as shown on Drawings.
_ Install sound barrier materials, gaskets, and clips as per manufacturers' instructions.
_ Do not allow any sound transfer connections within soundproof party wall construction.

E. Coordination quality control:

_ Do not allow trades to impair framing strength by cutting or drilling through members.
_ Do not change framing members without written consent of the Architect.
_ Provide fire protection facilities and all necessary fire protection precautions during construction.
_ Install required concealed fireproofing such as under enclosed stairs.
_ Provide openings for inspection of concealed work before closing in.

3.9 WOOD FRAMING -- BETWEEN PHASES AND AT CONCLUSION OF FRAMING

A. Inspection and cleanup:

_ Check and verify correctness of each stage of framing before installing subsequent framing:

_ Lumber grades, sizes, and spacing
_ Framing is plumb
_ Corners are square
_ Plates are level

_ Plates and cripples align vertically and horizontally

_ Remove all unusable wood scraps from site weekly and between each phase of framing.
_ Sweep work site clean weekly and between each phase of framing.
_ Do not bury any scraps or other trash on site.
_ Call for Architect and/or building department inspection before closing up concealed work.

3.10 FASTENERS, CONNECTORS, AND SUPPORTS -- INSTALLATION

A. Nailing and penetration:

_ Where not shown on nailing schedule, nails shall penetrate not less than 1/2 the length of nail.
_ Exception: 16d nails may connect two pieces of 2".
_ Nail at sufficient edge distance to avoid splitting wood.
_ Predrill as required.
_ Remove and replace split framing members.

_ Check nailing at each stage of framing before installing subsequent framing:

_ Quantities, spacing, and patterning as per building code.
_ Minimal bends.
_ Predrilled where required.
_ Nail heads flush or recessed as required.
_ Bent or used nails are not reused.

_ Use nailing machines or power hammers according to manufacturers' requirements.
_ Provide correct sizes and types of nails for use in nail guns.
_ Check and tighten all bolt connections after they're installed.
_ Recheck and retighten all bolt connections before final construction is completed.

B. Install joist hangers as per Drawings and manufacturer's instructions:

_ Spacings as per Drawings.
_ Set straight.
_ Aligned.
_ Completely secured at all connection points.
_ Secured with correct size and type fastenings.

C. Install bridging as per Drawings and manufacturer's instructions:

_ Placed so as to provide full bearing.
_ Set at joist midpoints or otherwise correctly spaced.
_ Bottoms are not nailed until the roof sheathing is laid.
_ Secured with correct size and type fastenings.

END OF SECTION

Notes:

_____

_____

_____

_____

DIVISION 6
WOOD AND PLASTIC

SAMPLE FINISH CARPENTRY CONSTRUCTION

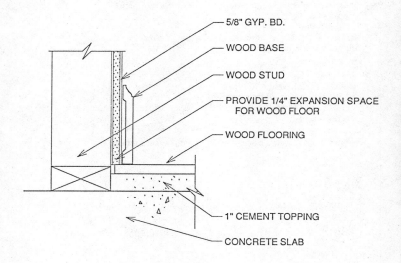

WOOD FLOOR @ BASE

DIVISION 6
WOOD AND PLASTIC

FINISH CARPENTRY AND MILLWORK
06200

PART 1 -- GENERAL

*Note to specifier: Include generic introductory PART 1 text from the beginning of Division 6. Then the MATERIALS text continues as follows:

PART 2 -- MATERIALS AND PRODUCTS

2.1 WOOD AND ACCESSORIES

A. Provide materials as per detail drawings, applicable trade standards, or approved samples:

_ Provide wood free of significant defects or deviations from grade standards.
_ Plywood and finish wood panels.
_ Fastenings and hardware.

_ Millwork materials:
_ Preservative treatment.
_ All-heart lumber.

_ Finish nails and putty.

PART 3 -- CONSTRUCTION AND INSTALLATION

3.1 PREPARATION

A. Tests and quality standards:

_ Reject any wood that is not certified as to grade.
_ Test and certify moisture content.
_ Provide all tests, certificates, and affidavits as required to verify quality of materials.
_ Do not have finish materials delivered until after the building is closed in.

_ Verify that special wood is marked and certified:
_ Preservative-treated.
_ Fire-resistant.
_ Kiln-dried.

_ Do not install finish panels with defects or deviations from grade standards:
_ Marred surfaces.
_ Cracks.
_ Defective patches.
_ Loose knots.
_ Split edges.
_ Delaminations.

B. Handling

_ Handle wood with care to avoid damage during:
_ Transport _ Unloading _ Moving _ Stacking

C. Storage:

_ Store wood as required to prevent damage and moisture absorption:
_ Off the floor.
_ Protected from weather.
_ Protected from all sources of water.
_ Neatly stacked to prevent warping.
_ Stacked with crosspieces for ventilation.
_ Secure from leaning or toppling.
_ In clean environment free of construction dust.

_ Properly ventilate wood treated with preservatives; store away from work areas.
_ Store kiln-dry materials to assure compliance with temperature and humidity restrictions.
_ Add preservative or backpriming to wood that will be exposed to potential rot conditions.

3.2 FINISH CARPENTRY -- EXTERIOR WORK

A. Protection:

_ Protect newly cut wood with prime coat or preservative treatment.
_ Protect with preservative, wood in contact with masonry or concrete.

_ Prepare siding surfaces for paint or stain:
_ Patch _ Sand _ Clean _ Preservatives _ Priming

B. Installation:

_ Install siding so that joints:
_ Are square.
_ Are staggered/patterned exactly as per Drawings.
_ Include expansion space at edges as required by manufacturers.

C. Cleanup

_ Remove all wood scraps from site.

3.3 CONSTRUCTION -- INTERIOR FINISH WORK

A. Coordination:

_ Coordinate with finish carpentry, furnishings, fixtures, and equipment to be installed by others:
_ Coordinate scheduling of deliveries and installation.
_ Coordinate backing materials and blocking.
_ Protect finish work from damage by other trades.

_ Prepare sub-surfaces to receive finish materials.

B. Working environment:

_ Keep working environment:
_ Clean, free of airborne construction dust.
_ Dry as required to maintain proper wood moisture content.
_ At comfortable working temperature.

C. Jointing:

   _ Make joints to conceal shrinkage:
      _ Miter exterior joints.
      _ Miter or scarf end-to-end joints.
      _ Cope interior joints.

_ All work per details and applicable trade standards:
_ Make saw cuts straight and clean.
_ Make tight fits, without gaps.
_ Make splices tight and staggered (never side by side).

_ Align and exactly match miter joints at edges and corners.
_ Install running trim in maximum lengths; do not use short pieces or splicing of scraps.

_ Wood joints:
_ Keep number of joints to a minimum by consistently using maximum size material.
_ Install tight joints without gaps.

_ Finish work:
_ Thoroughly sand smooth
_ Smooth edges.

D. Fastening:

_ Fasten all pieces straight, true, and secure.
_ Coordinate backing and blocking with other trades with interfacing work.
_ Nail exterior trim with galvanized nails.

E. Miscellaneous

_ Support shelves and closet poles so they will not sag when loaded.
_ Securely attach wood handrails that meet load tests.

F. Finishing:

_ Where sanding is required, sand with grain to totally smooth, unblemished surface.
_ Set finish nails before painting or staining.

_ Reject work as nonconforming due to:
_ Substandard material.
_ Hammer or other tool marks.
_ Dents.
_ Nailing splits.
_ Unmatched grains or patterns.
_ Uneven or over-sanding.

_ Re-sand any unfinished or unsmoothed surfaces.
_ Thoroughly clean and finish surfaces.
_ Protect finish work from construction damage.
_ Make repairs so they are undetectable.

G. Cleaning:

   _ Vacuum clean all work surfaces where sawdust
   _ Remove scraps frequently.
   _ Completely vacuum clean the work area upon completion of work.

END OF SECTION

Notes:

DIVISION 7

THERMAL AND MOISTURE PROTECTION
07000

CONTENTS

	INTRODUCTORY TEXT FOR ALL SECTIONS	169
07100	BELOW-GRADE WATERPROOFING	171
07100	FLOOR SLAB MEMBRANE DAMPPROOFING OR WATERPROOFING	173
07111	ELASTOMERIC MEMBRANE WATERPROOFING	175
07111	BITUMINOUS DAMPPROOFING	176
07200	BUILDING INSULATION	177
07311	ASPHALT SHINGLES	179
07313	WOOD SHINGLE ROOFING	181
07320	TILE ROOFING	183
07400	METAL ROOFING	185
07500	MEMBRANE ROOFING	187
07520	ASPHALT ROLL ROOFING	190
07530	SINGLE-PLY MEMBRANE ROOFING	192
07460	CEDAR SHINGLE SIDING	194
07560	HORIZONTAL WOOD SIDING	195
07600	FLASHING AND SHEET METAL	197
07631	GUTTERS AND DOWNSPOUTS	200
07800	EXPANSION AND CONTRACTION JOINTS	202
07900	SEALANTS	202
07800	SKYLIGHTS	206
07830	ROOF HATCHES	206

DIVISION 7
THERMAL AND MOISTURE PROTECTION

SAMPLE WATERPROOFING CONSTRUCTION

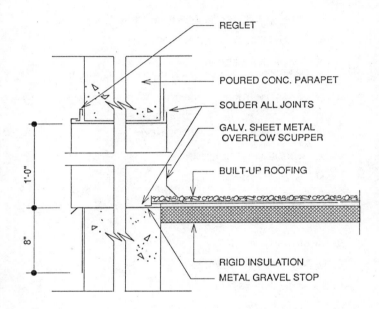

SCUPPER THROUGH PARAPET

DIVISION 7
THERMAL AND MOISTURE PROTECTION

INTRODUCTORY TEXT FOR ALL SECTIONS

*NOTE TO SPECIFIER: ADD SECTION TITLE AND CSI NUMBER.

PART 1   GENERAL

*NOTE TO SPECIFIER: THE FOLLOWING TEXT ON WORK, ETC.
IS INTRODUCTORY TEXT FOR ALL DIVISION 7 SECTIONS.

*Note to specifier: Whatever waterproofing, insulation, or roofing type you choose, you may wish to include some or all of the following generic text on Work, Quality Standards, Submittals, Materials Handling, and Preconstruction and Preparation at the start of each section.

1.1   WORK

A. Provide everything required to complete the work as shown on the Drawings and specified herein.

1.2   QUALITY STANDARDS

A. Provide experienced, well-trained workers competent to complete the work as specified.

B. Unless approved by the Architect, provide all related products and accessories from one manufacturer.

1.3   SUBMITTALS

A. Submit the following within ___ calendar days after receiving the Notice to Proceed.

*Note to specifier: Submittals are usually required within a specified number of calendar days after the Contractor is given the Notice to Proceed. 30 calendar days is a common requirement for medium- to large-size projects. Your choice of time will depend on the size of the project and the Owner's need for an expedited schedule.

_ Submit list of materials to be provided for this work.
_ Submit manufacturer's data required to prove compliance with these Specifications.
_ Submit manufacturer's installation instructions.
_ Submit shop drawings as required with complete details and assembly instructions.

*Note to specifier: Shop drawings may not be required if A/E drawings provide sufficient descriptive details for construction and installation.

_ Submit samples as required for approval by the Architect.

*Note to specifier: Samples may be requested with a different time limit than other submittals. Details of the samples requested -- sizes, finishes, etc. -- are usually specified.

1.4 MATERIALS HANDLING

A. Provide all materials required to complete the work as shown on drawings and specified herein.

_ Deliver, store, and transport materials to avoid damage to the product or to any other work.
_ Return any products or materials delivered in a damaged or unsatisfactory condition.
_ Materials and products delivered will be certified by the manufacturer to be as specified.
_ Keep all materials to be installed completely dry.
_ Keep insulation dry after installation and before covering with roofing.

_ Have on hand and ready for installation in coordination with roofing, all accessories such as:

_ Skylights   _ Hatches   _ Relief vents   _ Expansion joints

B. Store materials off the ground, protected from dirt, ground moisture, contaminants, and weather.

1.5 PRECONSTRUCTION AND PREPARATION

A. Examine and verify that job conditions are satisfactory for speedy and acceptable work.

_ Maintain and use all up-to-date construction documents on site.
_ Maintain and use up-to-date trade standards.
_ A preconstruction meeting will be held with all concerned parties if required by the Architect.
_ Confirm there are no conflicts between this work and prevailing building codes.
_ Notify Architect when work is scheduled to be installed.
_ Use agreed schedule for installation and for Architect's field observation.

*Note to specifier: Follow with PART 2 -- MATERIALS and PART 3 -- CONSTRUCTION AND INSTALLATION specifications as in the text that follows.

DIVISION 7
THERMAL AND MOISTURE PROTECTION

BELOW-GRADE WATERPROOFING
07100

PART 1 -- GENERAL

*Note to specifier: See note at the beginning of this Division on INTRODUCTORY TEXT FOR ALL SECTIONS.

PART 2 -- MATERIALS

*Note to specifier: Waterproofing of this type requires expert application and inspection. Use an independent consultant if you're not experienced with these products and rely on the consultant's specifications and inspection services. For smaller jobs, select a manufacturer with long experience with this work in your area and relay on the manufacturer's recommendations, specifications, and services.

2.1 WATERPROOFING

A. Provide and install:

*Note to specifier: Name product, trade name, and manufacturer(s).

PART 3 -- CONSTRUCTION AND INSTALLATION

*Note to specifier: Refer to manufacturer's instructions, trade association standards, and governing building code, as appropriate. See the note under PART 2 -- MATERIALS above.

3.1 APPLICATION

A. Strictly follow manufacturer's instructions to control application and guide inspection of the work.

　_ Provide certification of compliance as directed by the Architect.

B. Working conditions and surface:

_ Maintain work environment for best working conditions:
_ Dry   _ Above 50 degrees in temperature.
_ With enough room for worker safety and convenience.

_ Surface for application:
_ Smooth, without bumps, bulges, or holes   _ Clean   _ Dry
_ Clear of protruding materials

_ Add trowel coatings of mortar as required to:
_ Fill holes.
_ Fill areas of segregated aggregate.
_ Smooth out irregular wall surfaces.

C. Hollow masonry wall application:

_ Protect below-grade hollow masonry foundation walls:
_ Primer coat over clean, wet surface.
_ First coat scratched for bonding of second coat.
_ Second coat applied 24 hours after the first.
_ All coats kept damp for three-day curing.
_ Water repellent admixture/membranes added.

D. Cleaning, inspection, and repairs:

   _ Clean the work area and remove all scrap and excess materials
     from the site.
_ Allow convenient access for inspection of work.
_ Repair or replace defective work as directed by the Architect.

END OF SECTION

Notes:

DIVISION 7
THERMAL AND MOISTURE PROTECTION

FLOOR SLAB MEMBRANE DAMPPROOFING OR
WATERPROOFING
07100

PART 1 -- GENERAL

*Note to specifier: See note at the beginning of this Division on INTRODUCTORY TEXT FOR ALL SECTIONS.

PART 2 -- MATERIALS

*Note to specifier: Waterproofing of this type requires expert application and inspection. Use an independent consultant if you're not experienced with these products and rely on the consultant's specifications and inspection services. For smaller jobs, select a manufacturer with long experience with this work in your area and relay on the manufacturer's recommendations, specifications, and services.

2.1 WATERPROOFING

A. Provide and install: Manufactured by:

*Note to specifier: Name product, trade name, and manufacturer(s).

PART 3 -- CONSTRUCTION AND INSTALLATION

3.1 APPLICATION

*Note to specifier: Refer to manufacturer's instructions, trade association standards, and governing building code, as appropriate. See the note under PART 2 -- MATERIALS above.

A. Soil preparation:

_ Clear subgrade of weak soil.
_ Backfill with well compacted soil before laying gravel granular fill.
_ Cover granular fill thoroughly with grout coat as per Drawings.

B. Membranes:

_ Generously lap and thoroughly mop membrane felts over insulating slab or grout-covered gravel.
_ Immediately cover completed membrane with mortar coat.
_ Thoroughly lap and seal the joints of roll roofing used for dampproofing with mastic.
_ Waterproofing or dampproofing membranes must not be punctured or damaged in any way.

C. Membrane protection:

_ Provide complete protection to waterproofing or dampproofing membranes:
_ During laying of reinforcing _ Pour of slab

_ Don't pour concrete atop waterproofing or dampproofing membranes until they're inspected.
_ Verified membranes to be undamaged and sealed tight before pour.

D. Cleaning, inspection, and repairs:

_ Clean the work area and remove all scrap and excess materials from the site.
_ Allow convenient access for inspection of work.
_ Repair or replace defective work as directed by the Architect.

END OF SECTION

Notes:

DIVISION 7
THERMAL AND MOISTURE PROTECTION

ELASTOMERIC MEMBRANE WATERPROOFING
07111 (07110)

PART 1 -- GENERAL

*Note to specifier: See note at the beginning of this Division on INTRODUCTORY TEXT FOR ALL SECTIONS.

PART 2 -- MATERIALS

*Note to specifier: Waterproofing of this type requires expert application and inspection. Use an independent consultant if you're not experienced with these products and rely on the consultant's specifications and inspection services. For smaller jobs, select a manufacturer with long experience with this work in your area and relay on the manufacturer's recommendations, specifications, and services.

2.1 WATERPROOFING

A. Provide and install: Manufactured by:

*Note to specifier: Name product, trade name, and manufacturer(s).

PART 3 -- CONSTRUCTION AND INSTALLATION

3.1 APPLICATION:

A. Strictly follow manufacturer's instructions to control application and guide inspection of the work.

_ Provide certification of compliance as directed by the Architect.

*Note to specifier: See the note under PART 2 -- MATERIALS above.

B. Working conditions:

_ Examine work conditions to assure they are as required before application begins.
_ Determine the number of plies required by ground water hydrostatic head pressure.

C. Cleaning, inspection, and repairs:

   _ Clean the work area and remove all scrap and excess materials from the site.
_ Allow convenient access for inspection of work.
_ Repair or replace defective work as directed by the Architect.

END OF SECTION

DIVISION 7
THERMAL AND MOISTURE PROTECTION

BITUMINOUS DAMPPROOFING
07160

PART 1 -- GENERAL

*Note to specifier: See note at the beginning of this Division on INTRODUCTORY TEXT FOR ALL SECTIONS.

PART 2 -- MATERIALS

*Note to specifier: Waterproofing of this type requires expert application and inspection. Use an independent consultant if you're not experienced with these products and rely on the consultant's specifications and inspection services. For smaller jobs, select a manufacturer with long experience with this work in your area and relay on the manufacturer's recommendations, specifications, and services.

A. Provide and install: Manufactured by:

*Note to specifier: Name product, trade name, and manufacturer(s).

PART 3 -- CONSTRUCTION AND INSTALLATION

3.1 APPLICATION

A. Provide cold-applied asphalt bitumen dampproofing as shown on the Drawings and as specified herein.

B. Use materials that comply with the following standards:

_ Asphalt as per ASTM D449, Type I.
_ Asphalt primer as per ASTM D41.
_ Compatible with substrate.

C. Cleaning, inspection, and repairs:

_ Clean the work area and remove all scrap and excess materials from the site.
_ Allow convenient access for inspection of work.
_ Repair or replace defective work as directed by the Architect.

END OF SECTION

DIVISION 7
THERMAL AND MOISTURE PROTECTION

BUILDING INSULATION
07200

PART 1 -- GENERAL

*Note to specifier: See note at the beginning of this Division on INTRODUCTORY TEXT FOR ALL SECTIONS.

PART 2 -- MATERIALS

*Note to specifier: Select insulation to be used from the options below, and delete those not to be included. List manufacturer(s) and product names and the following data after your selection:

2.1 INSULATION

A. Insulation shall be (type):
   Manufactured by:

_ Location

_ Type (number)

_ R-value

_ Thickness

*Note to specifier: Select type from list below:

_ Glass fiber insulation:
_ Rigid
_ Semi-rigid

_ Batt insulation:
_ Preformed glass fiber batt.

_ Polystyrene insulation: Extruded cellular or molded bead type.

_ Glass fiber loose fill:
_ Fiberglass
_ Cellulose

B. Provide tapes, fastenings, and other related materials as instructed by insulation manufacturer.

PART 3 -- CONSTRUCTION AND INSTALLATION

3.1 PREPARATION AND MATERIALS HANDLING

A. Delivery:

_ Obtain manufacturer affidavit that materials delivered are as specified.

B. Storage:

_ Keep insulation materials totally dry at all times in storage and during installation.

3.2 APPLICATION

A. Preparation:

_ Keep areas to be insulated clean and dry.
_ Do not install insulation where it might be exposed to water.

B. Install as per manufacturer's instructions and building code requirements:

_ Install batts with tight contact of insulation with framing.
_ Use fastenings and fastening spacing as required by manufacturer.
_ Allow air spaces as required by manufacturer.
_ Keep rips and surface damage minimal; repair damage wherever it occurs.
_ Properly position reflective surface.
_ Cleanly cut and tightly fit batts around electrical and plumbing components.
_ Keep ventilation space unobstructed.

C. Loose insulation:

_ Do not allow air blower to over-fluff insulation.
_ Completely and evenly fill joist spaces to correct depth.
_ Thoroughly fill and pack all wall spaces.

D. Cleaning, inspection, and repairs:

_ Clean the work area and remove all scrap and excess materials from the site.
_ Allow convenient access for inspection of work.
_ Repair or replace defective work as directed by the Architect.

END OF SECTION

DIVISION 7
THERMAL AND MOISTURE PROTECTION

ASPHALT SHINGLES
07311

PART 1 -- GENERAL

*Note to specifier: See note at the beginning of this Division on INTRODUCTORY TEXT FOR ALL SECTIONS.

PART 2 -- MATERIALS

2.1 ASPHALT SHINGLES, ACCESSORIES, AND RELATED MATERIALS

A. Asphalt Shingles shall be as manufactured by:

_ Weight: lbs/square.

_ Tab type:

_ Manufacturer:

_ Product:

_ Color:

_ Warranty:

B. Accessories and related materials:

_ Underlayment: No. 15 asphalt-saturated roofing felt.

_ Nails:
_ Hot-dipped, zinc-coated steel.
_ Length sufficient to penetrate roof sheathing.

_ Staples:
_ Hot-dipped, zinc-coated steel.
_ Length sufficient to penetrate roof sheathing.

_ Plastic cement:
_ Asphalt-type cement with mineral fiber components.

_ Flashing:
_ Non-corrosive sheet metal.
_ Step flashing at all roof penetrations.

_ Drip edge at eaves and gables:
_ Non-corrosive sheet metal.

PART 3 -- CONSTRUCTION AND INSTALLATION

3.1 APPLICATION

A. Install granular-surfaced, asphalt shingle roofing as shown on the Drawings and as specified herein.

_ Do not install shingles when air temperature is lower than that required by the manufacturer.

B. Install as per manufacturer's instructions and in compliance with governing building code.

_ Install required preparatory flashing, and keep flashing to be installed later on hand.
_ Align starter course precisely.
_ Make certain layer and row alignment are correct.
_ Exposure will be as required by building code.
_ Spacing and nailing penetration will be as required by building code.
_ Work only when exterior air temperature is 40 degrees F. or higher.

C. Cleaning, inspection, and repairs:

_ Clean the work area and remove all scrap and excess materials from the site.
_ Allow convenient access for inspection of work.
_ Leave drains and other openings clear and clean of debris.
_ Provide walks or runways to protect roofing if there is to be continued construction work.
_ Repair or replace defective work as directed by the Architect.

END OF SECTION

Notes:

_____

_____

_____

_____

_____

_____

_____

_____

DIVISION 7
THERMAL AND MOISTURE PROTECTION

WOOD SHINGLE ROOFING
07313

*Note to specifier: Wood shingles and shakes are the worst source of residential fire and spread of fire to other buildings. Wood shingles cannot be used under any circumstances unless fully treated to retard fire in a manner approved by the governing building code.

PART 1 -- GENERAL

*Note to specifier: See note at the beginning of this Division on INTRODUCTORY TEXT FOR ALL SECTIONS.

PART 2 -- MATERIALS

2.1 WOOD SHINGLES AND ACCESSORIES

A. Provide wood shingle roofing as shown on the Drawings and as specified herein.

*Note to specifier: Make choice below and eliminate those options that are not applicable.

_ Materials shall be certified as fully fire retardant.

_ Red cedar shingles:
_ Number 1, Blue Label, medium weight.

_ Red cedar shakes:
_ Number 1, hand-split heavy shakes.

_ Fire rated wood shakes:
_ Hand-split Red Cedar heavy shakes, treated to provide Class B fire rating.

B. Accessories and related materials:

_ Underlayment:
_ No. 15 asphalt-saturated felt.

_ Nails:
_ Standard round wire shingle type.
_ Hot-dipped, zinc-coated steel.
_ Length sufficient to penetrate roof sheathing.

_ Staples:
_ Hot-dipped, zinc-coated steel.
_ Length sufficient to penetrate roof sheathing.

_ Flashing:
_ Non-corrosive sheet metal.
_ In icy climates, flash eaves with 30 lb. roofing felt or ice-shield membrane.

PART 3 -- CONSTRUCTION AND INSTALLATION

3.1 PREPARATIONS

A. Sheathing and related work prior to shingling:

*Note to specifier: Select the appropriate sheathing notes below:

_ Apply shingles over solid sheathing as shown on the Drawings.
_ Strip sheathing to be spaced at the same exposure as shingles.
_ Keep deck or sheathing dry prior to application.
_ Install deck or sheathing with firm and complete supports.
_ Install underlayment material and lapping as per manufacturer's instructions.
_ Install required hip, valley, and cricket flashings prior to shingling.
_ Install gutters that attach directly to the deck.
_ Broom-clean deck or sheathing.

3.2 INSTALLATION

A. Install as per manufacturer's instructions and in compliance with governing building code.

_ Space shingles 1/4" apart.
_ Space shakes 1/2" apart.
_ Space each joint a minimum of 1-1/2" from adjacent course.
_ Double up at first course to form a 1" drip edge.
_ Install shingles so there's no visible deviation from alignment.
_ Stagger joints as per manufacturer's instructions.
_ Shingle exposure to weather will be as required by code.
_ Nail spacing will be as required by building code.
_ Do not allow nails to penetrate decking or sheathing so as to be visible from below.

B. Related construction:

_ Provide related fire-retardant materials.
_ Provide compatible nails and other fastenings.
_ Provide compatible flashing materials.
_ Complete flashings as required during and at the conclusion of shingling.

C. Cleaning, inspection, and repairs:

_ Clean the work area and remove all scrap and excess materials from the site.
_ Allow convenient access for inspection of work.
_ Leave drains and other openings clear and clean of debris.
_ Provide walks or runways to protect roofing if there is to be continued construction work.
_ Repair or replace defective work as directed by the Architect.

END OF SECTION

DIVISION 7
THERMAL AND MOISTURE PROTECTION

TILE ROOFING
07320

PART 1 -- GENERAL

*Note to specifier: See note at the beginning of this Division on INTRODUCTORY TEXT FOR ALL SECTIONS.

PART 2 -- MATERIALS

2.1 TILE AND ACCESSORIES

A. Provide tile roofing as shown on the Drawings and as specified herein.

_ Roofing materials shall be:

_ Type:

_ Color:

_ Manufacturer:

B. Accessories and related materials:

_ Underlayment:

_ 30 lb. asphalt-saturated roofing felt.
_ Double felt at hips, valleys and ridges.

_ Trim for ridges, hips and gables must be as approved by tile manufacturer.

_ Wood stringers/strips:

_ Redwood or pressure treated.
_ Sized and applied as per tile manufacturer's instructions.

_ Fasteners:

_ Use large-headed copper or other non-corrosive nails.
_ 11 gauge.
_ Sufficient length for full penetration into sheathing.

_ Mortar:

_ One part cement to four parts sand.
_ Add color as recommended by tile manufacturer.

PART 3 -- CONSTRUCTION AND INSTALLATION

3.1 PREPARATION

A. Install base required for tile roofing as per building code requirements and manufacturer's instructions.

_ Roof deck: 1/2" sheathing minimum.
_ Battens: 1 x 2, nailed 24" o.c. with 8d galvanized nails.
_ Hips and ridges: 1 x 1, nailed 24" o.c. with 8d galvanized nails.
_ Install required preparatory flashing, and keep flashing to be installed later on hand.

3.2 INSTALLATION

A. Install roofing as per building code requirements and manufacturer's instructions.

_ Align starter course precisely.
_ Make certain layer and row alignment are correct.
_ Exposure will be as required by building code.
_ Spacing and nailing penetration will be as required by building code.

B. Cleaning, inspection, and repairs:

_ Clean the work area and remove all scrap and excess materials from the site.
_ Allow convenient access for inspection of work.
_ Leave drains and other openings clear and clean of debris.
_ Provide walks or runways to protect roofing if there is to be continued construction work.
_ Repair or replace defective work as directed by the Architect.

END OF SECTION

Notes:

_____
_____
_____
_____
_____
_____

DIVISION 7
THERMAL AND MOISTURE PROTECTION

METAL ROOFING
07400

PART 1 -- GENERAL

*Note to specifier: See note at the beginning of this Division on INTRODUCTORY TEXT FOR ALL SECTIONS.

PART 2 -- MATERIALS

2.1   METAL ROOFING

A. Provide metal roofing as shown on the Drawings and as specified herein.

*Note to specifier:

_ Roofing materials:

_ Type: Standing seam

              Batten

_ Material: Aluminum; thickness:

              Steel; thickness:

_ Manufacturer:

_ Finish/color:

B. Provide all related materials and accessories:

_ Roof deck:

_ APA -rated plywood as shown on the Drawings.

_ Underlayment:

_ 30 lb. asphalt-saturated roofing felt.

_ Fasteners only as approved by the manufacturer of the roofing product.

_ Trim manufactured or approved by the metal roofing manufacturer.

PART 3 -- CONSTRUCTION AND INSTALLATION

3.1 INSTALLATION

A. Install roofing as per building code requirements and manufacturer's instructions.

B. Cleaning, inspection, and repairs:

_ Clean the work area and remove all scrap and excess materials from the site.
_ Allow convenient access for inspection of work.
_ Leave drains and other openings clear and clean of debris.
_ Provide walks or runways to protect roofing if there is to be continued construction work.
_ Repair or replace defective work as directed by the Architect.

_ Touch-up:
_ Minor scratches and abrasions may be touched up.
_ Damaged material that may affect the integrity of the roofing must be replaced.

END OF SECTION

Notes:

DIVISION 7
THERMAL AND MOISTURE PROTECTION

MEMBRANE ROOFING
07500

PART 1 -- GENERAL

*Note to specifier: See note at the beginning of this Division on INTRODUCTORY TEXT FOR ALL SECTIONS.

PART 2 -- MATERIALS

*Note to specifier: Membrane roofing requires expert application and inspection. Use an independent consultant if you're not well experienced with these materials and rely on the consultant's specifications and inspection services. For smaller jobs, select a manufacturer with long experience with this work in your area and relay on the manufacturer's recommendations, specifications, and services.

2.1 MEMBRANE ROOFING

A. Provide and install:

Manufactured by:

*Note to specifier: Name product, trade name, and manufacturer(s).

PART 3 -- CONSTRUCTION AND INSTALLATION

3.1 MEMBRANE ROOF DECK PREPARATION

A. Deck construction:

_ Construct deck surface firm and fully supported.
_ Construct deck slopes so there will be no level areas or pockets that allow ponding.

B. Deck maintenance:

_ Keep deck surface smooth and free of irregularities.
_ Keep deck surface dry.
_ Cover voids in wood decks with sheet metal.
_ Broom-clean deck surface.
_ Dry out and vent insulating concrete deck.

C. Related construction:

_ Prime concrete walls and parapets adjacent to roof.
_ Stagger rigid insulation joints.
_ Do not construct insulation joints over metal deck joints or flutes.
_ Put in place and properly secure vapor retarder.
_ Secure vapor retarder to protect insulation at roof edge strips.
_ Install insulation vents in place.

_ Nailing and other fastenings for:
_ Louvers
_ Roof drains
_ Equipment supports

3.2 MEMBRANE ROOFING CONSTRUCTION

A. Nailing, lapping, and temperatures:

_ Do roof nailing as per roofing materials manufacturer's instructions.
_ Never allow felt laps to be less than the widths required by the manufacturer.
_ Roll or otherwise lay felts smoothly without pockets.
_ Keep felt at a temperature of no less than 50 degrees F.

_ Maintain bitumen at temperatures required by the roofing product manufacturer:
_ Correct kettle temperature without overheating.
_ Control heat loss during transport.
_ Correct temperature at application point.

B. Bitumen and mopping:

_ Provide sufficient quantities of bitumen for generous coverage of the felts.
_ Fasten and mop base sheets exactly as required by the manufacturer.
_ Mop felts over 100% of surface and do not allow felt-to-felt contact anywhere.
_ Mop felts completely over lap lines.
_ Pour the flood or final coat if recommended by the manufacturer.
_ Add a protective glaze coat if final coat and aggregate are delayed beyond the application period.
_ Apply aggregate on the hot surface final flood coat.

_ Combine and lap felts and bitumen with other materials as required by manufacturer:
_ Edge strips _ Flashing

C. Inspection and tests:

_ Test samples will be cut to verify layering, laps, and mopping.
_ When test samples are cut, complete thorough repair and patching promptly.
_ A roofing inspector may be provided at the discretion of the Owner.
_ Contractor shall provide free access for, and full cooperation with, any roofing inspector.

D. Cleaning and repairs:

_ Clean the work area and remove all scrap and excess materials from the site.
_ Allow convenient access for inspection of work.
_ Leave drains and other openings clear and clean of debris.
_ Provide walks or runways to protect roofing if there is to be continued construction work.
_ Repair or replace defective work as directed by the Architect.

END OF SECTION

Notes:

DIVISION 7
THERMAL AND MOISTURE PROTECTION

ASPHALT ROLL ROOFING
07520

PART 1 -- GENERAL

*Note to specifier: See note at the beginning of this Division on INTRODUCTORY TEXT FOR ALL SECTIONS.

PART 2 -- MATERIALS

2.1 ASPHALT ROLL ROOFING

A. Provide granular-surface asphalt roll roofing as shown on the Drawings and as specified herein.

_ Materials:
_ Mineral-surfaced asphalt roll roofing

_ Manufacturer:

_ Product:

_ Color:

2.2 RELATED MATERIALS AND ACCESSORIES

A. Provide as per manufacturer's instructions:

_ Accessories:
_ Underlayment: 15 lb. asphalt-saturated roofing felt.

_ Nails:
_ Hot-dipped, zinc-coated steel.
_ Length sufficient to penetrate roof sheathing.

_ Staples:
_ Hot-dipped, zinc-coated steel.
_ Length sufficient to penetrate roof sheathing.

_ Plastic cement:
_ Asphalt-type cement with mineral fiber components.

_ Flashing:
_ Non-corrosive sheet metal.
_ Step flashing at all roof penetrations.

_ Drip edge at eaves and gables:
_ Non-corrosive sheet metal.

PART 3 -- CONSTRUCTION AND INSTALLATION

3.1 PREPARATION

A. Preparation and substructure as per manufacturer's instructions and building code requirements.

3.2 INSTALLATION

A. Install as per manufacturer's instructions:

_ Side laps: 2" - 6" overlap   _ Blind nail at 9" o.c.
_ Exposed nails at 12" o.c.
_ End laps: 3" overlap
_ Nailed 3" o.c. minimum of 3' from laps on adjacent course.

B. Cleaning, inspection, and repairs:

_ Clean the work area and remove all scrap and excess materials from the site.
_ Allow convenient access for inspection of work.
_ Leave drains and other openings clear and clean of debris.
_ Provide walks or runways to protect roofing if there is to be continued construction work.
_ Repair or replace defective work as directed by the Architect.

END OF SECTION

Notes:

_____
_____
_____
_____
_____
_____
_____
_____
_____
_____
_____

DIVISION 7
THERMAL AND MOISTURE PROTECTION; ROOFING

SINGLE-PLY MEMBRANE ROOFING
07530

PART 1 -- GENERAL

*Note to specifier: See note at the beginning of this Division on INTRODUCTORY TEXT FOR ALL SECTIONS.

PART 2 -- MATERIALS

*Note to specifier: Membrane roofing requires expert application and inspection. Use an independent consultant if you're not well experienced with these materials and rely on the consultant's specifications and inspection services. For smaller jobs, select a manufacturer with long experience with this work in your area and relay on the manufacturer's recommendations, specifications, and services.

2.1 SINGLE PLY MEMBRANE ROOFING

A. Provide a fully-adhered elastomeric sheet membrane roofing system.

_ Roofing:

_ Material:

_ Color:

_ Thickness:

_ Manufacturer:

_ Substrate:

_ Provide smooth substrate surface of material approved by membrane manufacturer.
_ Fill surface joints and cracks that are wider than 1/4".

PART 3 -- CONSTRUCTION AND INSTALLATION

3.1   PREPARATION

A. Flashing must be as approved by membrane manufacturer.

_ Install approved flashing as per membrane manufacturers' instructions.
_ Install flashing at perimeter, curbs, skylights, vents, and all roof penetrations.

3.2 INSTALLATION

A. Install strictly as per manufacturer's instructions:

_ Lay membrane over substrate without stretching.
_ Lap edges a minimum of 3".
_ Allow membrane to rest one-half hour or more prior to splicing, flashing, and other work.
_ Adhere membrane directly to approved substrate without wrinkling.
_ Clean area of membrane to be seamed with approved cleaner.
_ Remove all visible talc and other contaminates.
_ Adhere seams with manufacturer-approved adhesive.
_ Avoiding globbing or puddling of adhesive.

B. Cleaning, inspection, and repairs:

_ Clean the work area and remove all scrap and excess materials from the site.
_ Allow convenient access for inspection of work.
_ Leave drains and other openings clear and clean of debris.
_ Provide walks or runways to protect roofing if there is to be continued construction work.
_ Repair or replace defective work as directed by the Architect.

END OF SECTION

Notes:

DIVISION 7
THERMAL AND MOISTURE PROTECTION

CEDAR SHINGLE SIDING
07460

PART 1 -- GENERAL

*Note to specifier: See note at the beginning of this Division on INTRODUCTORY TEXT FOR ALL SECTIONS.

PART 2 -- MATERIALS

2.1 WOOD SHINGLE SIDING

A. Provide shingle siding as shown on the Drawings:

*Note to specifier: Choose wood type and grade as appropriate to local custom, supply, and climate

Red cedar shingles: No. 1 Blue Label or No. 2 Red Label.
White cedar shingles: Clear grade.

B. Related materials and accessories:

_ Underlayment: 15 lb. asphalt-saturated roofing felt.

_ Fasteners:
    _ Use hot-dipped, zinc-coated nails   _ Stainless steel
    _ Aluminum nails or staples.
    _ Staples to be minimum 7/16", 16 gauge.
    _ Nails and staples of sufficient length to penetrate sheathing 1/2 - 3/4".

_ Flashing: Non-corrosive sheet metal.

PART 3 -- CONSTRUCTION AND INSTALLATION

3.1 APPLICATION

A. Sidewall application:

_ Double or triple shingles at foundation.
_ Shingles to be spaced 1/8" - 1/4" apart.
_ Joints to be minimum 1-1/2" from joints of adjacent course.
_ Exposure to be as shown on the Drawings.

*Note to specifier: Choose corner treatment options. Corner boards; size. Woven corners. Mitered corners.

B. Cleaning and repairs:

_ Clean the work area and remove all scrap and excess materials from the site.
_ Repair or replace defective work as directed by the Architect.

END OF SECTION

DIVISION 7
THERMAL AND MOISTURE PROTECTION

HORIZONTAL WOOD SIDING
07560

PART 1 -- GENERAL

*Note to specifier: See note at the beginning of this Division on INTRODUCTORY TEXT FOR ALL SECTIONS.

PART 2 -- MATERIALS

2.1 SIDING

A. Provide horizontal wood siding as shown on the Drawings.

*Note to specifier: Name the material and grade and fill in the following data as applicable:

_ Type (check all that apply to this project):

_ Bevel (clapboard); size: species:

_ Tongue & groove; size: species:

_ Shiplap; size: species:

_ Other:

B. Related materials and accessories:

_ Underlayment: Use 15 lb. asphalt-saturated roofing felt.

_ Fasteners:
_ Use hot-dipped, zinc-coated nails
_ Stainless steel nails
_ Aluminum nails

_ Nails and staples to be of sufficient length to penetrate sheathing 1/2 - 3/4".

_ Flashing: Non-corrosive sheet metal.

PART 3 -- CONSTRUCTION AND INSTALLATION

3.1 JOINTING AND NAILING

A. Nail siding firmly into studs.

_ All joints between siding pieces to be over studs.
_ Nails to be set and puttied.
_ Siding exposure: inches.

*Note to specifier: Name the siding exposure as appropriate to the material, grade, and climate. Follow the standards of the applicable wood working and grading association.

B. Primer:

_ Back prime all wood siding with approved primer.

C. Cleaning and repairs:

_ Clean the work area and remove all scrap and excess materials from the site.
_ Repair or replace defective work as directed by the Architect.

END OF SECTION

Notes:

DIVISION 7
THERMAL AND MOISTURE PROTECTION

FLASHING AND SHEET METAL
07600

PART 1 -- GENERAL

*Note to specifier: See note at the beginning of this Division on INTRODUCTORY TEXT FOR ALL SECTIONS.

PART 2 -- MATERIALS

*Note to specifier: Flashing requires expert application and inspection. Construction details require extra care and should be carefully checked and coordinated with these Specifications.

2.1 FLASHING

A. Provide and install all flashing of types, sizes and gauges shown in the Drawings and specified herein:

B. Provide flashing materials as follows:

TYPE	LOCATION	MATERIAL	MANUFACTURER

C. Flashing and sheet metal in this Section includes but is not limited to:

_ Cap flashing _ Stepped flashing _ Through-wall flashing
_ Edge flashing _ Hip flashing _ Ridge flashing _ Valley flashing
_ Crickets
_ Gutters, scuppers, and downspouts
_ Flashing at doors and windows

D. Unless noted otherwise on drawings, gauges and standards for flashing materials shall be:

_ Steel: 20 gauge galvanized steel, G90 galvanizing, ASTM A525.
_ Copper: 16 oz./sq.ft. cold-rolled copper, ASTM B370.
_ Aluminum: 20 gauge alloy 3003 clear anodized aluminum, ASTM B209.
_ PVC: 30 mil. sheet.

_ Laminated copper/fabric flashing for masonry flashing:
_ 3 oz. copper sheet laminated between 2 sheets of bituminous saturated fabric.

PART 3 -- CONSTRUCTION AND INSTALLATION

*Note to specifier: Refer to manufacturer's instructions, trade association standards, and governing building code, as appropriate. See the note under PART 2 -- MATERIALS above.

3.1 APPLICATION

A. Strictly follow manufacturer's instructions.

B. Provide flashing connections and fabrications as detailed:

_ Cleating _ Soldering _ Welding _ Bolting
_ Riveting _ Anchorage
_ Use non-corrosive fastenings.
_ Keep dissimilar metals well separated to avoid corrosion.
_ Lap and lock seams; solder seam joints where necessary to guarantee watertightness.
_ Lap edge metal at least 4".
_ Install flashing inserts in walls deeply as detailed, secured and caulked.

C. Roof flashing

_ Integrate and embed edge flashing within roofing membrane as detailed.
_ Apply and mop additional plies of felt as detailed and as per manufacturer's instructions.
_ Provide and install flashing with widths and laps as detailed.
_ Caulk and paint hip, ridge, and other exposed flashing.
_ Cover all edges of metal laps with adhesive.
_ Fill joints between flashing and the edges of shingles with adhesive.
_ Caulk all reglets.

_ Provide flashing, cement, and caulking for all roof accessories:

_ Roof hatches _ Vents _ Curbs _ Louvers
_ Roof drains _ Equipment supports

D. Cleaning and repairs:

_ Clean the work area and remove all scrap and excess materials from the site.
_ Repair or replace defective work as directed by the Architect.

END OF SECTION

Notes:

_____

_____

_____

DIVISION 7
THERMAL AND MOISTURE PROTECTION

SAMPLE GRAVEL STOP CONSTRUCTION

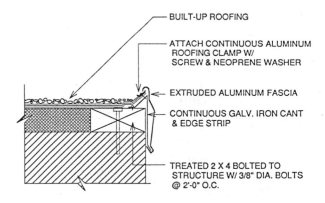

GRAVEL STOP- FORMED

DIVISION 7
THERMAL AND MOISTURE PROTECTION

GUTTERS AND DOWNSPOUTS
07631

PART 1 -- GENERAL

*Note to specifier: See note at the beginning of this Division on INTRODUCTORY TEXT FOR ALL SECTIONS.

PART 2 -- MATERIALS

*Note to specifier: See materials notes in the preceding section on FLASHING AND SHEET METAL. In addition, you may make reference to: Aluminum, painted: 0.040" thick.
Aluminum, anodized: 0.050" thick.

PART 3 - CONSTRUCTION AND INSTALLATION

3.1 APPLICATION

A. Strictly follow manufacturer's instructions.

B. Provide connections and fabrications as detailed.

C. Install to provide ample support and proper drainage as follows:

_ Use hangers and straps adequate in size and spacing to support loads.
_ Support every separate section.
_ Construct gutters with positive slopes, to prevent accumulation of standing water.
_ Install downspouts that are visually plumb and that match specified tolerances.
_ Lap joints to match drainage flow.
_ Provide at least one expansion/contraction joint midway between each gutter downspout.
_ Provide movement slip joints on downspouts.
_ Protect building surfaces from damage from hanger and strap connectors.
_ Provide screens, strainers, and covers, to prevent debris from accumulating in drains.
_ Keep downspouts and gutters separated from wall surfaces to avoid staining and corrosion.

D. Cleaning and repairs:

_ Clean the work area and remove all scrap and excess materials from the site.
_ Leave drains clear, clear, and free of debris.
_ Repair or replace defective work as directed by the Architect.

END OF SECTION

DIVISION 7
THERMAL AND MOISTURE PROTECTION

SAMPLE MOVEMENT JOINT CONSTRUCTION

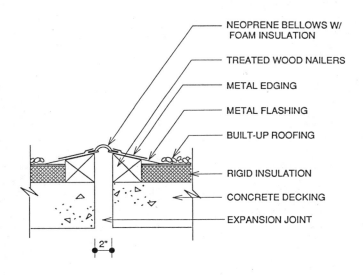

ROOF EXPANSION JOINT

DIVISION 7
THERMAL AND MOISTURE PROTECTION

EXPANSION AND CONTRACTION JOINTS
07800

SEALANTS
07900

PART 1 -- GENERAL

*Note to specifier: See note at the beginning of this Division on INTRODUCTORY TEXT FOR ALL SECTIONS.

PART 2 -- MATERIALS

*Note to specifier: This work is crucial for failures prevention and the long-term maintenance of the building. Selected materials from the most reputable manufacturers and use them exactly according to their recommended applications and procedures. Use the manufacturer's specifications, and adapt them to this format.

2.1 SEALANTS

A. Provide sealants and related materials as manufactured by:

*Note to specifier: List types, locations, and applications of sealants. See the note above under PART 2 -- MATERIALS on the importance of relying on manufacturer's specifications and instructions.

*Note to specifier: List specific products and manufacturers for:

_ Sealant material:

_ Sealant application location:

_ Primer:

_ Backup materials:

_ Bond-breaker:

_ Masking tape:

_ Colors:

_ Deliver compounds in sealed, labeled containers.

*Note to specifier: Refer to manufacturer's literature, and choose sealants according to the following qualities required for the work:

_ Plus-or-minus movement capacity.
_ Recovery.
_ Adhesive strength to joined material; cohesive strength.
_ Life expectancy.
_ Curing time.
_ Shrinkage.
_ Stability, flexibility, hardness.
_ Water and chemical resistance.

PART 3 -- CONSTRUCTION AND INSTALLATION

3.1 APPLICATION

A. Preparation:

_ Construct vertical and horizontal joints at locations and sizes shown in the Drawings.
_ Verify each joint type according to the Drawings.
_ Verify depth and width of each joint.
_ Verify the backing and sealant material required for each joint.
_ Uniformly apply solvents required to prepare surfaces; apply with one rag, remove with another.
_ Remove dust and moisture after sandblasting or wire brushing of concrete and masonry.
_ Carefully mask around joints where sealant might discolor or stain finish surfaces.
_ Protect extra-wide joints from accumulation of debris and from damage to sealant.

B. Preparing materials to receive sealant:

_ Concrete and tile surfaces receiving sealant shall be clean, dry, and well brushed.
_ Steel surfaces receiving sealant shall be sandblasted or wire brushed as required for bonding.
_ Aluminum receiving sealant will be completely clean of protective coatings, dirt, and oil.
_ Cleaners used on materials to receive sealant shall be as approved by sealant manufacturer.

C. Application of backup material:

_ Clean mortar or other debris from movement joints prior to application of backing and sealant.
_ Use only tools recommended or supplied by the sealant manufacturer.
_ Pack deep joint slots with backup material, to bring joints to correct depth.
_ Place backup material by compressing 25% to 50% to a tight, complete fit in the joint space.

D. Primer and bond breaker:

_ Use only a primer approved by the sealant manufacturer.
_ Apply exactly as instructed by the primer manufacturer.
_ Use only a bond breaker approved or manufactured by the sealant manufacturer.
_ Apply exactly as instructed by the primer manufacturer.

E. Application of sealant:

_ Apply sealant under pressure as required to completely fill the joints.
_ Carefully mask around joints where sealant might discolor or stain finish surfaces.
_ Tool joints to a smooth, consistent profile.

3.2 CLEANUP AND REPAIR

A. Cleanup:

   _ Remove masking immediately after joints are tooled.
   _ Clean adjacent surfaces as instructed by the sealant manufacturer.
   _ Remove all debris and empty containers from the job site.

B. Repairs:

   _ Repair or reseal defective work as directed by the Architect.

END OF SECTION

Notes:

_____
_____
_____
_____
_____
_____
_____
_____
_____
_____

DIVISION 7
THERMAL AND MOISTURE PROTECTION

SAMPLE FLASHING & REGLET CONSTRUCTION

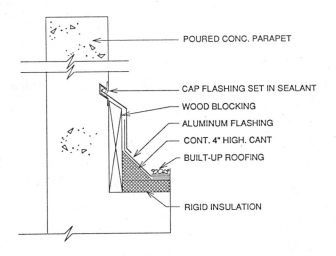

FLASHING @ CONC. PARAPET

DIVISION 7
THERMAL AND MOISTURE PROTECTION

SKYLIGHTS
07800

ROOF HATCHES
07830

PART 1 -- GENERAL

*Note to specifier: See note at the beginning of this Division on INTRODUCTORY TEXT FOR ALL SECTIONS.

PART 2 -- MATERIALS AND PRODUCTS

2.1 GENERAL

A. Product shall be as manufactured by:

*Note to specifier: Include note "(Product) shall be as manufactured by (company name)." As appropriate, add product type, trade name, grade, size, and manufacturer's location.

PART 3 -- CONSTRUCTION AND INSTALLATION

3.1 WORK CONDITIONS

A. Correct any conditions not in compliance with Section 1.5.A. noted above.

B. Correct any conditions that might interfere with speedy, well coordinated execution of the work.

C. All work conditions shall be as per:

_ Manufacturer's instructions.
_ Trade association standards.
_ Governing building and safety codes.

3.2 PREPARATION

*Note to specifier: Follow manufacturer, trade association, or consultant's specifications for this subsection.

3.3 INSTALLATION

A. Install products as per Drawings and these Specifications.

*Note to specifier: Follow manufacturer, trade association, or consultant's specifications for this subsection.

B. Upon completion:

_ Secure all required tests, inspections, and approvals of the completed system.
_ Make all required adjustments and corrections at no added cost to the Owner.

C. Provide for maintenance of this work for one year following final acceptance by Owner.

_ Maintenance includes all work required in manufacturer's instructions:
    _ Inspection and adjustment.
    _ Repair and replacement of parts as required.

3.4 REPAIR AND CLEANUP

A. After installation, inspect all work for improper installation or damage.

_ Operating hardware must perform smoothly.
_ Repair or replace any defective work.
_ Repair work will be undetectable.
_ Redo repairs if work is still defective, as directed by the Architect.

B. Clean the work area and remove all scrap and excess materials from the site.

END OF SECTION

Notes:

_____

_____

_____

_____

_____

_____

_____

_____

_____

_____

_____

Notes:

DIVISION 8

DOORS AND WINDOWS
08000

CONTENTS

08100 METAL DOORS AND FRAMES	211
08200 WOOD DOORS	215
08250 PACKAGED DOOR ASSEMBLIES	219
08300 SPECIAL DOORS	219
08410 ENTRANCES AND STOREFRONTS	221
08500 METAL FRAME WINDOWS	224
08600 WOOD FRAME WINDOWS	227
08710 HARDWARE	229

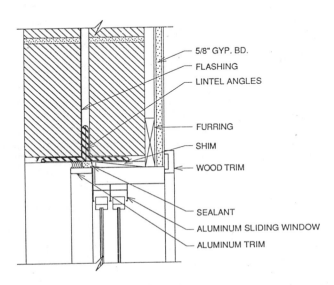

ALUM. SLIDING -- HEAD @ BLOCK WALL

DIVISION 8
DOORS AND WINDOWS

SAMPLE METAL THRESHOLDS

FULL-SCALE GENERIC DETAILS

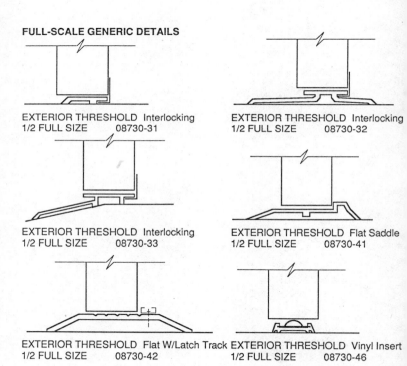

EXTERIOR THRESHOLD Interlocking
1/2 FULL SIZE   08730-31

EXTERIOR THRESHOLD Interlocking
1/2 FULL SIZE   08730-32

EXTERIOR THRESHOLD Interlocking
1/2 FULL SIZE   08730-33

EXTERIOR THRESHOLD Flat Saddle
1/2 FULL SIZE   08730-41

EXTERIOR THRESHOLD Flat W/Latch Track
1/2 FULL SIZE   08730-42

EXTERIOR THRESHOLD Vinyl Insert
1/2 FULL SIZE   08730-46

DIVISION 8
DOORS AND WINDOWS

METAL DOORS AND FRAMES
08100

PART 1 -- GENERAL

1.1 WORK

A. Provide metal doors and frames where shown on the Drawings and as specified herein.

B. Door and frame types and sizes shall be as per the Drawings and Door Schedule.

PART 2 -- MATERIALS AND PRODUCTS

2.1 METAL DOORS

A. Provide full flush doors as per Drawings and Door Schedule.

*Note to Specifier: The most common metal door requirements are as follows:

_ 18 gauge metal for interior doors.
_ 16 gauge metal for exterior doors.
_ Reinforced for finish hardware.

B. Provide doors that are straight, free of defects and blemishes, and that have:

_ Smooth edges.
_ Clean joints.
_ Consistent, blemish-free finish.
_ Correct finish material thickness.

C. Doors will be complete with reinforcing and backing plates:

_ Lock blocks, other blocks required for hardware.
_ Stile edges and astragals for paired doors.

D. Verify that factory preparation and prefitting follow required hardware templates:

_ Cuts   _ Drilling   _ Routing   _ Accessories

E. Provide door glazing with:

_ Stops as required.
_ Labeled safety glass.

F. Provide fire-rated doors that comply with all building code and fire code requirements.

_ Openings/glass are as required.
_ Hinges are as required.
_ Undercut does not exceed the allowable code maximum.
_ Correct identification labels and/or certification.
_ Wire glass lights.
_ Fusible links for louvers.
_ Correct factory-applied hardware.
_ Automatic closure hardware.

G. Doors shall be as manufactured by:

*Note to Specifier: Specify acceptable options in manufactured metal doors.

2.2 METAL FRAMES

A. Provide metal frames as per Drawings and Door Schedule.

*Note to Specifier: The most common metal door frame requirements are as follows:

_ Welded frames with mitered corners.
_ 16 gauge metal for interior doors.
_ 14 gauge frames for interior door frames over 5' wide.
_ 14 gauge metal for exterior doors.
_ Reinforced for finish hardware.
_ Doors and frames shall be straight, free of defects and blemishes, with:
_ Smooth edges   _ Clean joints
_ Consistent, blemish-free finishes
_ Correct finish material thickness

METAL DOORS AND FRAMES

B. Metal frames shall be as manufactured by:

*Note to Specifier: Specify acceptable choice or options in manufactured metal frames, or leave to the discretion of the Contractor.

C. Provide cleaned, shop-primed frames ready for finish painting.

_ Painting as per Section 09900 of these Specifications.

2.3 FINISH HARDWARE

A. Manufacturer shall prepare frames for finish hardware using hardware supplier's templates.

B. Use hardware supplier's templates to install or prepare for all finish hardware.

PART 3 -- CONSTRUCTION AND INSTALLATION

3.1 PRECONSTRUCTION AND PREPARATION

A. Examine and verify that job conditions are satisfactory for speedy and acceptable work.

_ Maintain and use all up-to-date construction documents on site.
_ Maintain and use up-to-date trade standards.

B. Do not allow door swings to conflict with:

_ Electrical switches or outlets. _ Wall guards or rails.

3.2 INSTALLATION

A. Mounting frames:

_ Mount frames prior to wall construction wherever practical to do so.
_ Mount frames plumb, straight, and securely braced until permanently anchored.
_ Repair any damage caused by temporary bracing.
_ Mount frames in place with screws and anchors, only as approved by manufacturer.
_ Block for hinges and other hardware.
_ Grout frames during construction.
_ Caulk frames for complete seal.

B. Installing doors:

_ Schedule door installation to avoid construction damage.

_ Hang doors:
_ Straight _ Level _ Plumb _ Smooth in opening
_ Smooth and secure in closing

_ Provide clearances below doors as necessary to allow for:
_ Thresholds _ Weatherstripping
_ Gasketing _ Carpet

C. Cutting and sealing doors:

_ Do not cut fire-rated doors so as to negate fire rating.
_ Seal or reseal doors whenever they are cut.
_ Seal doors at tops and bottoms after installation.
_ Apply first coats of paint or sealer to both sides of doors during the same painting operation.
_ Seal, stain, or paint exterior doors before or immediately after installing them.
_ Treat exterior doors before or immediately after they're installed.

3.3 INSPECTION, REPAIR, AND TOUCH-UP

A. After installation, inspect all doors and frames to find and repair damaged surfaces.

_ Repair or replace any damaged materials as directed by the Architect.
_ Repair or replace any other materials damaged during installation.
_ Sand rusted and damaged surfaces smooth, and touch up prime coat with compatible primer.
_ Make undetectable repairs before applying final finish.
_ Repair or replace any related damaged or non-complying materials.
_ Repair or replace doors that do not operate freely and smoothly.
_ Replace any doors that are not in compliance with Specifications.
_ Any costs for replacing doors for non-compliance will be paid by the Contractor.

B. Final door mounts:

_ Square.
_ Smooth operating.
_ Plumb when doors are closed, partially open, and fully open.

END OF SECTION

Notes:

DIVISION 8
DOORS AND WINDOWS

WOOD DOORS
08200

PART 1 -- GENERAL

1.1 WORK

A. Provide wood doors, complete with hardware where shown on the Drawings and as specified herein.

B. Provide and install all door hardware as shown on schedules and as specified herein.

1.2 QUALITY STANDARDS

A. Provide experienced, well-trained workers competent to complete the work as specified.

B. Unless approved by the Architect, provide all related products and accessories from one manufacturer.

C. Comply with standards of the Architectural Woodwork Institute for the grades specified.

1.3 SUBMITTALS

A. Submit the following within _____ calendar days after receiving the Notice to Proceed.

*Note to specifier: Submittals are usually required within a specified number of calendar days after the Contractor is given the Notice to Proceed. 30 calendar days is a common requirement for medium- to large-size projects. Your choice of time will depend on the size of the project and the Owner's need for an expedited schedule.

_ Submit list of materials to be provided for this work.
_ Submit manufacturer's data required to prove compliance with these Specifications.
_ Submit manufacturer's installation instructions.
_ Submit 8" x 8" samples of each specified door face material.

1.4 MATERIALS HANDLING

A. Provide all materials required to complete the work.

_ Deliver and transport materials to avoid damage to the product or to any other work.
_ Return any products or materials delivered in a damaged or unsatisfactory condition.
_ Materials and products delivered will be certified by the manufacturer to be as specified.

B. Delivery:

_ Deliver after interior finish materials are dry.
_ Deliver after building reaches average long-term interior humidity.
_ Packaging must be sealed with clear manufacturer and identification markings.
_ Provide manufacturer's representative's certification that doors delivered are as specified.
_ Seal all edges of unfinished doors.

C. Storage:

_ Store flat on 2x4's spaced at 12" centers.
_ Provide sheet materials at bottom and top sides; to protect doors from damage.
_ Lift and carry doors when moving them, do not drag into position.

_ Store delivered doors:
_ Consistently vertical or flat.
_ Supported off floor.
_ Protected from weather and moisture.
_ Protected from construction damage.

PART 2 -- MATERIALS AND PRODUCTS

2.1   08200   WOOD DOORS

A. Provide wood doors as per Drawings and Door Schedule.

_ Solid or hollow-core as shown on Door Schedule.
_ UL labels as shown on Door Schedule.

B. Doors shall be manufactured by:

*Note to Specifier: Specify acceptable manufacturers of wood doors.

C. Provide doors that are straight, free of defects and blemishes, and that have:

_ Smooth edges.
_ Clean joints.
_ Consistent, blemish-free finishes.
_ Correct finish material thickness.

D. Verify that factory preparation and prefitting follow required hardware templates:

_ Cuts   _ Drilling   _ Routing   _ Accessories
_ Hollow-core doors must have core construction as required to receive finish hardware.

E. Provide door glazing with:

_ Stops as required.
_ Labeled safety glass.

F. Provide fire-rated doors that comply with all building code and fire code requirements.

_ Openings/glass are as required.
_ Hinges are as required.
_ Undercut does not exceed the allowable code maximum.
_ Correct identification labels and/or certification.
_ Wire glass lights.
_ Fusible links for louvers.
_ Correct factory-applied hardware.
_ Automatic closure hardware.

G. Wood construction and finish for wood doors shall be as shown in Door Schedule.

*Note to specifier, check notes in Door Schedule re: _
    Grade _   Species _
    Veneer cut/match   Edge banded

PART 3 -- INSTALLATION

3.1 PRECONSTRUCTION AND PREPARATION

A. Examine and verify that job conditions are satisfactory for speedy and acceptable work.

_ Maintain and use all up-to-date construction documents on site.
_ Maintain and use up-to-date trade standards.

B. Do not allow door swings to conflict with:

_ Electrical switches or outlets.
_ Wall guards or rails.

3.2 INSTALLATION

A. Mount frames and doors plumb, straight, and securely braced.

B. Mounting tolerances:

_ Bottom clearance, 1/2" max.
_ Top clearance, 1/8" max.
_ Lock and hinge edge, bevel at 1/8" in 2" max.

C. Installing doors:

_ Schedule door installation to avoid construction damage.

_ Hang doors:
_ Straight _ Level _ Plumb _ Smooth in opening
_ Smooth and secure in closing

_ Provide clearances below doors as necessary to allow for:
_ Thresholds _ Weatherstripping _ Gasketing _ Carpet

_ Install fastenings and hardware as per Hardware Schedule and instructions of manufacturer.

D. Cutting and sealing doors:

_ Do not cut fir-rated doors so as to negate fire rating.
_ Seal or reseal doors whenever they are cut.
_ Seal doors at tops and bottoms after installation.
_ Apply first coats of paint or sealer to both sides of doors during the same painting operation.
_ Seal, stain, or paint exterior doors before or immediately after installing them.
_ Treat exterior doors before or immediately after they're installed.

3.3 INSPECTION, REPAIR, AND TOUCH-UP

A. After installation, inspect all doors and frames to find and repair damaged surfaces.

_ Repair or replace any damaged materials as directed by the Architect.
_ Repair or replace any other materials damaged during installation.
_ Make undetectable repairs before applying final finish.
_ Repair or replace any related damaged or non-complying materials.
_ Repair or replace doors that do not operate freely and smoothly.
_ Replace any doors that are not in compliance with Specifications.
_ Any costs for replacing doors for non-compliance will be paid by the Contractor.

B. Final door mounts:

_ Square.
_ Smooth operating.
_ Plumb when doors are closed, partially open, and fully open.

END OF SECTION

Notes:

_____
_____
_____
_____
_____
_____
_____

DIVISION 8
DOORS AND WINDOWS

PACKAGED DOOR OPENING ASSEMBLIES
08250

*Note to specifier: List special door opening equipment specified in this section. Use the generic three-part format, GENERAL, MATERIALS, and CONSTRUCTION/INSTALLATION common to the other specifications sections. Refer to architectural drawings and Shop Drawings. Refer to specific manufacturer or choices of manufacturers of special doors, and follow manufacturers' recommended specifications.

SPECIAL DOORS
08300

*Note to specifier: List special doors specified in this section. Use the generic three-part format, GENERAL, MATERIALS, and CONSTRUCTION/INSTALLATION common to the other specifications sections. Refer to architectural drawings and Shop Drawings. Refer to specific manufacturer or choices of manufacturers of special doors, and follow manufacturers' recommended specifications.

DIVISION 8
DOORS AND WINDOWS

SAMPLE WOOD DOOR FRAME CONSTRUCTION

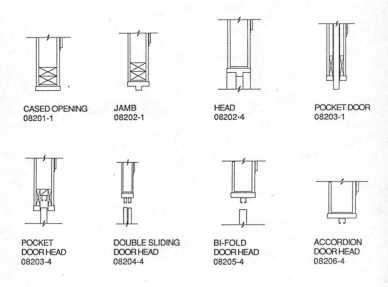

CASED OPENING
08201-1

JAMB
08202-1

HEAD
08202-4

POCKET DOOR
08203-1

POCKET
DOOR HEAD
08203-4

DOUBLE SLIDING
DOOR HEAD
08204-4

BI-FOLD
DOOR HEAD
08205-4

ACCORDION
DOOR HEAD
08206-4

DIVISION 8
DOORS AND WINDOWS

ENTRANCES AND STOREFRONTS
08410

PART 1 -- GENERAL

1.1 WORK

*Note to specifier: Refer to architectural drawings and Shop Drawings. Refer to specific manufacturer or choices of manufacturers of special doors, and follow manufacturers' recommended specifications.

A. Provide metal frame storefronts and entrances where shown on the Drawings and as specified herein.

B. Frame types and sizes shall be as per the Drawings and Window Schedule.

1.2 QUALITY STANDARDS

A. Provide experienced, well-trained workers competent to complete the work as specified.

B. Unless approved by the Architect, provide all products from one manufacturer.

1.3 SUBMITTALS

A. Submit the following within _____ calendar days after receiving the Notice to Proceed:

_ Submit list of materials to be provided for this work.
_ Submit manufacturer's data required to prove compliance with these Specifications.
_ Submit manufacturer's installation instructions.
_ Submit Shop Drawings with complete details and assembly instructions.

*Note to specifier: Submittals are usually required within a specified number of calendar days after the Contractor is given the Notice to Proceed. 30 calendar days is a common requirement for medium- to large-size projects. Your choice of time will depend on the size of the project and the Owner's need for an expedited schedule.

1.4 SAMPLES

A. Submit a sample of each frame component.

_ Samples should show the maximum range of light to dark and color difference.
_ Approved samples will be retained by the Architect to verify that final work matches samples.

B. Provide all materials required to complete the work as shown on Drawings and specified herein.

_ Deliver and transport materials to avoid damage to the product or to any other work.
_ Return any products or materials delivered in a damaged or unsatisfactory condition.
_ Materials and products delivered will be certified by the manufacturer to be as specified.

1.5 MATERIALS HANDLING

A. Provide all materials required to complete the work as shown on Drawings and specified herein.

_ Deliver and transport materials to avoid damage to the product or to any other work.
_ Return any products or materials delivered in a damaged or unsatisfactory condition.
_ Materials and products delivered will be certified by the manufacturer to be as specified.

B. Store materials safely to avoid damage, and locate to expedite the work.

1.6 WARRANTY

A. Provide to the Owner a warranty guaranteeing to replace defective work.

_ Time limit of the warranty is ___ years from the Date of Substantial Completion.

*Note to Specifier: Three years is a common warranty period for this construction.

_ Defective work includes but is not limited to:

_ Failure of, or difficulty in operation of movable components.
_ Air infiltration.
_ Visibly defective finish.
_ Any other defects relevant to appearance, operation, or potential failure.

PART 2 -- MATERIALS AND PRODUCTS

2.1 ENTRANCES AND STOREFRONTS

A. Provide storefronts complete with glazing as per Drawings and Window Schedule.

B. Storefront and entrance assemblies shall be as manufactured by:

MATERIAL	GAUGES	FINISHES

Curtain wall framing:

Main entrance doors:

Secondary entrance doors:

Flashing:

*Note to specifier: Note whether storefront frames are to be of steel, aluminum, or bronze, and the type of finish. Include reference to the automatic door operating system if included as part of this work.

2.2 FINISH HARDWARE

A. Provide hardware as per Section 08710 and related sections of these Specifications.

_ Fit non-surface-mounted finish hardware at factory.
_ Use concealed fasteners wherever possible.

PART 3 -- ENTRANCES AND STOREFRONTS CONSTRUCTION AND INSTALLATION

3.1 PRECONSTRUCTION

A. Examine and verify that job conditions are satisfactory for speedy and acceptable work.

B. Store materials safely to avoid damage, and locate to expedite the work.

3.2 INSTALLATION

*Note to specifier: Refer to architectural drawings and Shop Drawings. Refer to specific manufacturer or choices of manufacturers of special doors, and follow manufacturers' recommended specifications.

*Note to specifier: Your specifications should be based on recommendations of your chosen manufacturer(s) but also see clauses in the sections in this Division on GLASS AND GLAZING.

END OF SECTION

DIVISION 8
DOORS AND WINDOWS

METAL FRAME WINDOWS
08500

PART 1 -- GENERAL

1.1 WORK

A. Provide metal frame windows where shown on the Drawings and as specified herein.

B. Window types and sizes shall be as per the Drawings and Window Schedule.

*Note to specifier: Note whether windows are of steel, aluminum, bronze, etc.

1.2 QUALITY STANDARDS

A. Provide experienced, well-trained workers competent to complete the work as specified.

B. Unless approved by the Architect, provide all related products and accessories from one manufacturer.

1.3 SUBMITTALS

A. Submit the following within _____ calendar days after receiving the Notice to Proceed:

_ Submit list of materials to be provided for this work.
_ Submit manufacturer's data required to prove compliance with these Specifications.
_ Submit manufacturer's installation instructions.

*Note to specifier: Submittals are usually required within a specified number of calendar days after the Contractor is given the Notice to Proceed. 30 calendar days is a common requirement for medium- to large-size projects. Your choice of time will depend on the size of the project and the Owner's need for an expedited schedule.

1.4 MATERIALS HANDLING

A. Provide all materials required to complete the work as shown on Drawings and specified herein.

_ Deliver and transport materials to avoid damage to the product or to any other work.
_ Return any products or materials delivered in a damaged or unsatisfactory condition.
_ Materials and products delivered will be certified by the manufacturer to be as specified.

B. Store materials safely to avoid damage and located to expedite the work.

PART 2 -- MATERIALS AND PRODUCTS

2.1 WINDOWS

A. Provide windows complete with glazing as per Drawings and Window Schedule.

B. Windows shall be as manufactured by:

*Note to Specifier: List manufacturer or choices of manufacturers with model names or numbers, finishes, and all required related descriptive information.

PART 3 -- CONSTRUCTION AND INSTALLATION

3.1 PRECONSTRUCTION

A. Examine and verify that job conditions are satisfactory for speedy and acceptable work.

B. Store materials safely to avoid damage and located to expedite the work.

3.2 INSTALLATION

*Note to specifier: Refer to architectural drawings and Shop Drawings. Refer to specific manufacturer or choices of manufacturers of windows, and follow manufacturers' recommended specifications.

A. Window dimensions and alignments shall be as per Drawings and Window Schedule.

_ Heads and sills are at correct elevations
_ Heads and sills level
_ Jambs at locations dimensioned on plan.
_ Jambs plumb and aligned with other facade elements

B. Installing windows:

_ Install and tightly secure frame braces and anchors.
_ Install sash free of bends, warps, and dents.
_ Protect aluminum frames and other materials especially vulnerable to damage.
_ Protect materials from contact with mortar or other sources of staining and chemical damage.
_ Place windows into openings without forcing.
_ Do not use windows or ventilators at any time as supports or rests for ladders and scaffolds.
_ Prepare weep holes adequate in size and spacing.
_ Keep weep holes clean.
_ Install windows that are weathertight and that allow no air infiltration.
_ Install ventilator hardware to operate easily and without sticking.
_ Use closures that are uniform and tight when units are closed and locked.
_ Install operable windows that open and close smoothly, without rattling or sticking.

C. Tolerances:

_ Construct openings of six feet or less within plus or minus 1/16 inch tolerance in each direction.
_ Construct openings larger than six feet within plus or minus 1/8 inch tolerance in each direction.
_ Construct openings with diagonal dimensions within 1/8 inch of each other.

3.3 INSPECTION, REPAIR, AND TOUCH-UP

A. After installation, inspect all windows and frames to find and repair damage.

_ Repair or replace any damaged materials as directed by the Architect.
_ Repair or replace any other materials damaged during installation.
_ Make undetectable repairs before applying final finish.
_ Repair or replace any related damaged or non-complying materials.
_ Repair or replace windows that do not operate freely and smoothly.
_ Replace any windows that are not in compliance with Specifications.
_ Any costs for replacing windows for non-compliance will be paid by the Contractor.

B. Final windows mounted:

   _ Square   _ Smooth operating

C. Protection and repair:

_ Provide barriers or clear separations between dissimilar metals.
_ Protect shop coat treatment from damage.
_ Touch up damaged shop coats before final painting.
_ Keep hardware clean.
_ After installation, protect finishes from physical and chemical damage.
_ Clean and protect metal and glass after installation.
_ Make undetectable repairs to damaged materials or finishes.

END OF SECTION

Notes:

DIVISION 8
DOORS AND WINDOWS

WOOD FRAME WINDOWS
08600

*Note to specifier: Adapt text from 06500, METAL WINDOWS. Specifications text, except for the references to metal or aluminum windows, will be identical to 06500. Refer to architectural drawings and Shop Drawings. Refer to specific manufacturer or choices of manufacturers of wood windows, and follow manufacturers' recommended specifications.

**DIVISION 8
DOORS AND WINDOWS**

SAMPLE WINDOW FRAME CONSTRUCTION

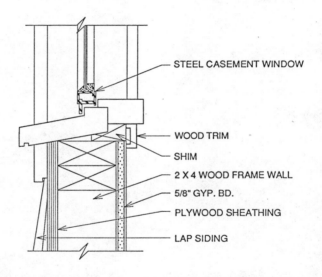

CASEMENT SILL @ WOOD STUD WALL

DIVISION 8
DOORS AND WINDOWS

HARDWARE
08710

*Note to specifier: Hardware requirements for larger projects are so intricate that you may need to hire an architectural hardware consultant to complete the hardware schedule and prepare the Specifications. Otherwise, rely on a selected hardware manufacturer's literature, and write a generic specification, as follows:

PART 1 -- GENERAL

1.1 WORK

A. Provide and install finish hardware where shown on the Drawings and as specified herein.

B. Provide and deliver finish hardware as required to be factory installed.

1.2 QUALITY STANDARDS

A. Provide experienced, well-trained workers competent to complete the work as specified.

B. Provide a member of the American Society of Architectural Hardware Consultants:

  _ To consult with the Architect during construction.
  _ Inspect all installed finish hardware.

1.3 SUBMITTALS

A. Submit the following within _____ calendar days after receiving the Notice to Proceed.

*Note to specifier: Submittals are usually required within a specified number of calendar days after the Contractor is given the Notice to Proceed. 30 calendar days is a common requirement for medium- to large-size projects. Your choice of time will depend on the size of the project and the Owner's need for an expedited schedule.

  _ Submit list of materials to be provided for this work.
  _ Submit manufacturer's data required to prove compliance with these Specifications.
  _ Submit manufacturer's installation instructions.
  _ Deliver templates and hardware requiring factory installation to the door manufacturer.

1.4 SAMPLES

A. Submit a sample of each hardware item within _____ calendar days after request by the Architect.

B. Samples will be returned to the Contractor for installation as specified.

1.5 MATERIALS HANDLING

A. Provide all materials required to complete the work as shown on Drawings and specified herein.

_ Deliver and transport materials to avoid damage to the product or to any other work.
_ Return any products or materials delivered in a damaged or unsatisfactory condition.
_ Packaging must be sealed with clear manufacturer and identification markings.
_ Materials and products delivered will be certified by the manufacturer to be as specified.

B. Store materials safely to avoid damage, and locate to expedite the work.

PART 2 -- MATERIALS AND PRODUCTS

2.1 HARDWARE

A. Provide hardware as per Hardware Schedule.

_ Provide hardware with unblemished finishes.
_ Match up hardware pieces in finish and color when they are used together or in close proximity.
_ Provide all fastenings and auxiliary devices necessary to complete the work.

B. Hardware shall be as manufactured by:

*Note to Specifier: Refer to schedules that follow or list manufacturer or choices of manufacturers with model names or numbers, finishes, and all required related descriptive information.

C. Provide these hardware groups in quantities as shown on the Drawings:

*Note to Specifier: Provide a complete list of hardware groups. Typical items for door hardware are shown below under "ITEM." Work for a larger project or building with extraordinary security requirements will require expert consultation.

HARDWARE GROUP 1

ITEM   FINISH OR CATALOG #
MANUFACTURER

Hinge
Lock or latch
Pull
Pushbar
Closer
Stop
Threshold
Astragal
Head and jamb seals
Weatherstrip
(*Other)

HARDWARE GROUP 2

ITEM   FINISH OR CATALOG #
MANUFACTURER

Hinge
Lock or latch
Pull
Pushbar
Closer
Stop
Threshold
Astragal
Head and jamb seals
Weatherstrip
(*Other)

*Continue as required with HARDWARE GROUP 3, etc.

2.2 KEYING

A. Provide complete locks and key system as shown on Hardware Schedule and specified herein.

B. Provide factory key and masterkey locks and cylinders.

_ Provide locks from a single manufacturer unless approved otherwise by the Architect.
_ Provide finishes as shown on the Hardware Schedule and as per approved samples.

*Note to Specifier: Fill in quantities below:

_ Number of keys for each lock:
_ Number of masterkeys:
_ Number of grand masterkeys:

C. Provide construction masterkey system.

- Upon Substantial Completion, change all construction locks, and install finish keying.
- Factory stamp keys: DO NOT DUPLICATE.
- Tag permanent keys, and provide certified delivery to the Owner.
- Provide a complete set of tools and maintenance manuals for all locks and operable hardware.

PART 3 -- CONSTRUCTION AND INSTALLATION

3.1 PRECONSTRUCTION

A. Examine and verify that job conditions are satisfactory for speedy and acceptable work.

B. Delivery, storage, coordination:

- Store materials safely to avoid damage and located to expedite the work.
- Store hardware securely, and provide exactly as needed for orderly completion of the work.
- Coordinate with all related trades and work, to expedite hardware installation.

3.2 INSTALLATION REQUIREMENTS -- GENERAL

A. Temporarily remove or cover exposed hardware when painting or cleaning adjacent materials.

B. Attach all hardware:

- Securely.
- With fasteners made specifically for that hardware.
- With fasteners suited to door or door frame materials and construction.
- Without damage to hardware or fasteners.

C. Match hardware:

- Match in type, size, and finish, all sets of fastenings, such as screws on hinge butts.
- Match all required screws to all screw-attached hardware, such as hinges.

D. Setting:

- Set all flush-set hardware such as hinge butts so they are truly flush without any protrusion.

E. Install doors to open and close:

- Easily, without binding.
- Quietly.
- With secure fit at latches.
- With tight fit at frames.

3.3 INSTALLATION -- BUTTS AND HINGES

A. Provide and install butts and hinges as shown in Hardware Schedule and as specified herein.

B. Hardware shall be as manufactured by:

*Note to Specifier: List manufacturer or choices of manufacturers with model names or numbers, finishes, and all required related descriptive information.

C. Install as per manufacturer's instructions:

_ Mortise-type hinges are flush.
_ Mortise edge distances on door are correct.
_ Top and bottom hinge heights are correct.
_ Space intermediate hinges equidistant from others.
_ Use 1/2 surface hinges on composite doors.

_ Screws:
_ Complete in number   _ Flush   _ Straight
_ Set tight without damaging screw slots.

D. Completely set in pins on butts.

E. Where butt hinges will swing 180 degrees, use hinges with adequate throw to clear the door trim.

3.4   INSTALLATION -- LOCKSETS, LATCHES, AND KEYS

A. Install locksets as per manufacturer's instructions:

_ Jig bore or predrilling.
_ Mortise for strike allows latchbolt to project fully.
_ Install cylinder cores with tumblers set upward.

_ Backsets:
_ Straight
_ Clear stops

_ Carefully guard master keys during construction.
_ Remove construction locks, and install permanent locks.
_ Match up locks and tagged keys after installing all units.
_ Verify that all keys and locks operate smoothly without effort.

B. Deliver keys and instructions to owner:

_ With permanent tags
_ Checked with door locks
_ Indexed
_ With key cabinet

3.5 INSTALLATION -- DOOR CLOSERS

A. Install per manufacturer's instructions, with special attachments as required for:

_ Wood doors _ Metal doors
_ Install closer fasteners straight, true, and undamaged.

B. After adjustment, verify that door closers operate:

_ Smoothly at correct speed
_ Without noise.
_ Firmly to close and latch doors.
_ Verify that door arms of closers are straight out when doors are closed.

3.6 INSTALLATION -- PLATES, DOOR STOPS, AND HOLDERS

A. Install door stops as per manufacturer's instructions:

_ Secure firmly to solid blocking or backing.
_ Allow proper door opening clearance.

B. Install stops of correct type and in correct positions to fully protect:

_ Adjacent surfaces _ Trim _ Hardware _ Furnishings.

C. Install push-, pull-, and kickplates according to door manufacturers instructions:

_ At correct heights from finish floor.
_ Within correct edge distances.
_ On correct sides of doors.
_ Firmly attached with fasteners suited to door construction.

3.7 INSTALLATION -- PANIC BARS AND FLUSH BOLTS

A. Install exit hardware as per manufacturer's instructions:

_ Exit cross bars: _ Level   _ Both arms securely attached.
_ Move simultaneously when pressed and released.

_ Top and bottom bolts:
_ Seat properly.
_ Are straight.
_ Stabilized as necessary as with mullion stabilizers.
_ Have correct depth of bolt throw in strike.

3.8 INSTALLATION -- MISCELLANEOUS HARDWARE AND ACCESSORIES

A. Smoke- or heat-sensing magnetic door holding devices:

_ Fire code approved and certified.
_ With prewired, dedicated electrical outlets.

B. Install hanging, folding, bifold door hardware as per approved samples.

C. Install suspended and hanging door tracks:

_ Level.
_ Match to weight of door panels.
_ Secure tightly.
_ Secure into supportive framing.
_ Clean and clear of obstructions.

D. Install sliding door tracks:

_ Level.
_ Secure tightly.
_ Secure into supportive framing.
_ Clean and clear of obstructions.

E. Adjust sliding door wheels for:

_ Smooth level glide.
_ Easy opening.
_ Secure closure of doors into latches.

F. Provide and install acoustic strips:

_ Shall be firmly attached.
_ Provide a secure sound seal.
_ Do not allow strips to interfere with door operation.

G. Install weatherstripping as per manufacturer's instructions:

_ Match samples.
_ Install after setting, adjusting, and checking doors.
_ Fasten securely.
_ Set to not scrape finish flooring.
_ Set to not interfere with door operation.

H. Cleaning and repair:

_ Keep hardware clean.
_ After installation, protect finishes from physical and chemical damage.
_ Clean and protect all hardware as recommended by manufacturers.
_ Replace or make undetectable repairs to damaged materials or finishes.

END OF SECTION

Notes:

_____

_____

_____

Notes:

DIVISION 9

FINISHES
09000

CONTENTS

09110	METAL STUD WALLS AND PARTITIONS	243
09120	SUSPENSION CEILING SYSTEMS	247
09200	LATH AND PLASTER	251
09215	VENEER PLASTER	256
09250	GYPSUM WALLBOARD	258
09300	CERAMIC TILE	264
09300	QUARRY TILE	264
09400	TERRAZZO	269
09450	INTERIOR STONE	271
09500	ACOUSTICAL INSULATION	272
09550	WOOD FLOORING	273
09550	WOOD FLOORING -- PARQUET AND BLOCK	273
09660	ASPHALT, VINYL AND RESILIENT TILE	278
09650	RESILIENT SHEET FLOORING	281
09680	CARPET	283
09900	PAINTING	286
09950	WALL COVERINGS	294

DIVISION 9
FINISHES

SAMPLE WOOD FLOOR CONSTRUCTION

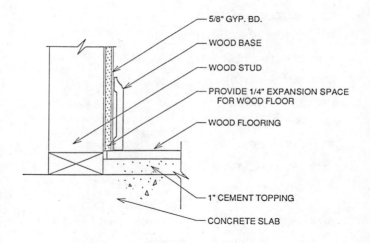

WOOD FLOOR @ BASE

DIVISION 9
FINISHES

*Note to specifier: This is a generic PART 1 text that applies to all Sections in this Division. Center the title and CSI number of the Specification Section in this title space.

PART 1 -- GENERAL

1.1 WORK

A. Provide:

*Note to specifier: Name the product, material, or finish such as any of the items listed on the preceding page. Since these are all highly specialized materials and systems, you will most likely need to use selected manufacturer's specifications or those from a consultant. You can modify those specifications to include the standard introductory format such as provided below.

B. Provide everything required to complete the work as shown on the Drawings and specified herein.

C. Other related work:

*Note to specifier: This would include reference to related construction that is not provided by the product or system manufacturer such as adjacent substrate preparation.

1.2 QUALITY STANDARDS

A. Provide experienced, well-trained workers competent to complete the work as specified.

B. Unless approved by the Architect, provide all related products and accessories from one manufacturer.

C. Use products and accessories:

_ From a manufacturer who specializes in making, installing, and servicing systems of this type.
_ From a manufacturer specified or approved by the Architect.

D. All work shall comply with manufacturer's instructions and governing building and safety codes.

*Note to specifier: Include reference to relevant trade standards, if there are any.

1.3 SUBMITTALS

A. Submit the following within _____ calendar days after receiving the Notice to Proceed.

*Note to specifier: Submittals are usually required within a specified number of calendar days after the Contractor is given the Notice to Proceed. 30 calendar days is a common requirement for medium- to large-size projects. Your choice of time will depend on the size of the project and the Owner's need for an expedited schedule.

_ Submit list of materials to be provided for this work.
_ Submit manufacturer's specifications required to prove compliance with these specifications.
_ Submit manufacturer's installation instructions.
_ Submit Shop Drawings as required with complete details and assembly instructions.
_ Submit Shop Drawings showing relationship and interface with adjacent or related work.
_ Submit samples of proposed exposed finishes and hardware for approval by the Architect.

*Note to specifier: Samples may be requested with a different time limit than other submittals. Details of the samples requested -- sizes, finishes, etc. -- are usually specified.

1.4 MATERIALS HANDLING

A. Provide all materials required to complete the work as shown on Drawings and specified herein.

_ Deliver, store, and transport materials to avoid damage to the product or to any other work.
_ Return any products or materials delivered in a damaged or unsatisfactory condition.
_ Materials and products delivered will be certified by the manufacturer to be as specified.

B. Store materials in a safe, secure location, protected from dirt, moisture, contaminants, and weather.

1.5 PRECONSTRUCTION AND PREPARATION

A. Examine and verify that job conditions are satisfactory for speedy and acceptable work.

_ Maintain and use all up-to-date construction documents on site.
_ Maintain and use up-to-date trade standards and manufacturer's instructions.
_ Confirm there is no conflict between this work and governing building and safety codes.
_ Confirm there are no conflicts between this work and work of other trades.
_ Confirm that work of other trades that must precede this work has been completed.
_ Meet all requirements to secure any applicable warranty.

B. Planning and coordination.

_ Notify Architect when work is scheduled to be started and completed.
_ If required by the Architect, a preconstruction meeting will be held with all concerned parties.
_ Use agreed schedule for installation and for field observation by Architect.

Notes:

**DIVISION 9
FINISHES**

SAMPLE SUSPENDED CEILING CONSTRUCTION
09110

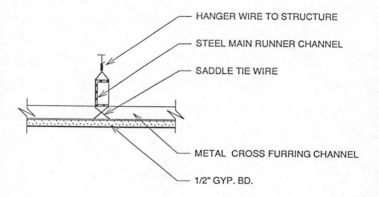

SUSPENDED GYPSUM BOARD CEILING
SCALE 3"=1'-0"

DIVISION 9
FINISHES

METAL STUD WALLS AND PARTITIONS
09110

PART 1 -- GENERAL

*Note to specifier: Include generic introductory PART 1 text from the beginning of Division 9.

PART 2 -- MATERIALS

2.1 METAL STUDS, CHANNELS AND ACCESSORIES

A. Metal studs and accessories shall be manufactured by:

*Note to specifier: Include note "(product) shall be manufactured by (company name). As appropriate, add product type, trade name, grade, size, and manufacturer's location.

B. Steel as per ASTM A446, A568, or A611.

_ 20 gauge steel studs and runners for load-bearing walls and
    reinforcement for door frames.
_ 18 gauge steel studs for non-load-bearing partitions.
_ 14 gauge "C" studs for exterior studs.
_ Steel studs and channels as per Federal specification QQ-S-775
    Type 1, Class E.

*Note to specifier: Metal stud gauges and strength of members vary from manufacturer to manufacturer so substitutions must be treated with caution and are allowable only where alternate materials completely match specified standards.

C. Galvanized coating as per ASTM A525.

D. Grout as required for leveling floor runners.

PART 3 -- CONSTRUCTION AND INSTALLATION

3.1 WORK CONDITIONS AND COORDINATION

A. Examine the job site, and correct any conditions that might interfere
    with speedy and acceptable work.

B. Coordination with other work -- utilities, fixtures, equipment, finishes:

_ Coordinate electrical stub-ups with the wall and partition framing plan.
_ Align floor-mounted electric outlet boxes with finish wall lines.
_ Coordinate stud walls with plumbing:
_ Supply lines _ Floor drains _ Thru-building roof drains

_ Do not allow HVAC ducts in wall framing to protrude beyond face of framing.
_ Supply and coordinate in-wall fixture and equipment supports:
_ Anchors _ Brackets _ Grounds _ Chairs _ Frames

_ Provide in-wall blocking, anchors, brackets, grounds, and other supports for wall-supported:
_ Plumbing fixtures _ Electrical fixtures
_ HVAC equipment
_ Kitchen or shop equipment _ Bathroom accessories
_ Handrails
_ Guards/protective rails _ Shelves _ Storage units
_ Fire hose cabinets

3.2 INSTALLATION

A. Install studs of the sizes and spacings shown on the Drawings.

B. Install as per manufacturer's instructions, applicable trade standards, and governing building code.

_ Provide solid continuous support under floor runners.
_ Stud spacings as required by manufacturer to firmly support facing materials.
_ Align holes for attached materials.
_ Install heavy gauge studs and extra stiffeners at jambs.
_ Install reinforced/heavy gauge studs at stress points.
_ Provide hot-dip galvanized metal where subject to moisture or fumes.
_ Provide cushioned space at top plate at slab to allow for slab deflection/movement.
_ Level tops of partitions for alignment with ceilings or top slab construction.

C. Provide backing and anchors for:

_ Door frames _ Opening frames _ Wall-mounted fixtures
_ Wall-mounted counters and cabinets _ Shelves _ Lockers
_ Recessed fixtures
_ Attached equipment _ Rails and grab bars
_ Toilet room partitions _ Chair rails
_ Acoustical separation for noisy fixtures and equipment

D. Tolerances:

_ Make framing surfaces flush within a tolerance of 1/8" in 10' in any direction.
_ Align and plumb studs to a tolerance of 1/8" per 10' horizontally and vertically.
_ Make top plate level within 1/8" per 10 linear feet.

E. Sound attenuation:

_ Set runners on continuous beads of sealant or other cushion as per
   manufacturer's instructions.
_ Stagger with plate separations for soundproofing.
_ Caulk for soundproofing.

3.3 CLEANUP AND REPAIR

A. After installation, inspect all work for improper installation or damage.

_ Repair or replace any work damaged during installation.
_ Redo defective repairs at the direction of the Architect.
_ Repair work will be undetectable.

END OF SECTION

Notes:

**DIVISION 9
FINISHES**

SAMPLE SUSPENDED CEILING CONSTRUCTION

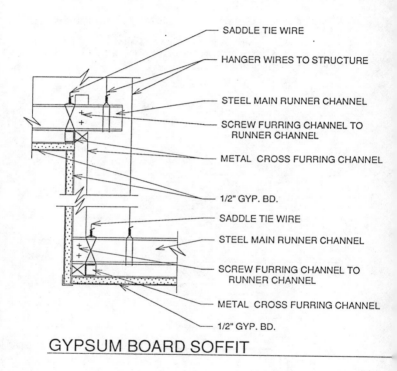

GYPSUM BOARD SOFFIT

DIVISION 9
FINISHES

SUSPENDED CEILING SYSTEMS
09120

PART 1 -- GENERAL

*Note to specifier: Include generic introductory PART 1 text from the beginning of Division 9.

PART 2 -- MATERIALS AND ACCESSORIES

2.1 GRID SYSTEM

A. Provide a complete grid system of light fixtures and grilles as per Drawings.

_ Include all required supports, anchors, and other accessories for a complete ceiling system.
_ Provide materials and finishes as per samples approved by the Architect.
_ All systems must be approved by the Underwriters Laboratories, Inc.
_ Systems must comply with governing building code and safety codes.

B. Provide and install products manufactured by:

*Note to specifier: Name manufacturer, product name, trade name, and all other data necessary to clearly identify the specified product.

_ Ceiling system type 1:

_ Ceiling system type 2:

_ Ceiling system type 3:

_ Ceiling system type 4:

_ Equal products of other manufacturers may be submitted for approval by the Architect.

2.2 ACOUSTICAL CEILING PANELS

A. Provide and install ceiling panels manufactured by:

*Note to specifier: Name manufacturer, product name, trade name, and all other data necessary to clearly identify the specified product.

_ Ceiling system type 1:

_ Ceiling system type 2:

_ Ceiling system type 3:

_ Ceiling system type 4:

_ Equal products of other manufacturers may be submitted for approval by the Architect.

B. Acoustical panels:

_ Type: 3/4" thick, molded.

*Note to specifier: Designate panel size and pattern such as 24" x 48" or 24" x 24", medium fissured mineral tile.

_ Panel edge: Square edge; flush mount with grid.
_ Panel edge: Reveal edge; panel extends below grid.
_ Align pattern in same direction.
_ Panels: 5/8" thick, felted

*Note to specifier: Designate square edge and flush mount. Refer to manufacturer's literature to make your selection.

_ Panel size: 24" by 48".
_ Panel edge: Square edge; flush mount with grid.
_ Align pattern in same direction.

2.3 CEILING TILES

A. Provide and install acoustical tiles manufactured by:

*Note to specifier: Designate tile size and pattern such as 3/4" x 12" x 12" or 24" x 24", medium natural fissured mineral tile.

B. Exposed grid suspension system:

_ Intermediate-duty painted steel.

*Note to specifier: Select concealed grid system with concealed spline or T-Grid exposed system.

2.4 OTHER MATERIALS AND RELATED WORK

A. Provide all related materials and accessories required for the work.

_ Channels and hangers as detailed, including:
_ Ties _ Clips _ Anchors

_ Special attachments as required:
_ Extra intermediate hanger wire _ Sway bracing
_ Turnbuckles _ Isolators
_ Hold-down clips on fire-rated ceilings

_ Through-ceiling fire barriers as required.
_ Through-ceiling sound walls as required.
_ Additional sound treatment above ceiling as directed by the Architect.

2.5 ADDITIONAL MAINTENANCE SUPPLY OF MATERIALS

A. Deliver an extra supply of 10% or more of each type of acoustical material installed.

_ Package each material, mark it clearly, and protect it against damage and deterioration.

PART 3 -- CONSTRUCTION AND INSTALLATION

3.1 PREPARATION

A. Preparation and coordination:

_ Measure and layout ceiling to avoid panels less than 1/2 panel in size.
_ Keep and use up-to-date trade standards on the job site.
_ Keep and use correct manufacturer's instructions.
_ Follow ceiling layout to provide clearances for all other work.

3.2 INSTALLATION

A. Install ceiling system as per manufacturer's instructions, ASTM C636 and governing building code.

_ Firmly attach hangers with special anchors as required for support by superstructure above:
_ Follow ceiling grid space pattern as designed in ceiling plans or approved by the Architect.
_ Provide hanger ties -- three twists per tie -- close to each anchor.
_ Do not allow loads of anything but the ceiling system on the ceiling supports.
_ Do not allow ceiling to brace or support mechanical or electrical equipment.
_ Label ceiling maintenance and inspection access panels.
_ Adjust ceiling: _ Level _ Aligned to walls

B. Provide fire and sound barriers as shown on Drawings.

_ Provide undamaged, uninterrupted, and tightly sealed through-ceiling fire barriers.
_ Install through-ceiling sound walls undamaged and tightly sealed.
_ Sound treatment above ceiling as directed by the Architect.
_ Do not allow paint applied to tiles or boards to affect acoustical properties.
_ Any additional required connectors and fastenings must match the rest of ceiling.

C. Bracing:

_ Provide lateral bracing for secure installation and as per governing building code.
_ Firmly secure lateral bracing to structural members.
_ Secure lateral bracing at right angles to partition.
_ Secure bracing four ways in large ceiling areas.
_ Provide hold-down clips for ceiling boards.

D. Tolerances:

_ Level whole ceiling within a tolerance of 1/4" in 10' and straight within the same tolerance.

*Note to specifier: 1/8" in 10' tolerance is possible if ceiling is low or otherwise highly visible.

3.3 CLEANING AND REPAIR

A. Cleaning:

_ Keep work area thoroughly clean and remove all scrap daily as work proceeds.

B. Repair:

_ Repair or replace defective work.
_ Make repairs undetectable.

END OF SECTION

Notes:

_____

_____

_____

_____

DIVISION 9
FINISHES

LATH AND PLASTER
09200

PART 1 -- GENERAL

*Note to specifier: Include generic introductory PART 1 text from the beginning of Division 9.

PART 2 -- MATERIALS

2.1 FURRING, LATH, AND ACCESSORIES

A. Metal lath and accessories shall be manufactured by:

*Note to Specifier: If gypsum lath is to be used, revise this text as per specifications of the gypsum lath manufacturer.

*Note to specifier: Include note "(product) shall be manufactured by (company name). As appropriate, add product type, trade name, grade, size, and manufacturer's location.

B. Lath:

_ Metal lath as per Federal Specification QQ-L-101.
_ Paper-backed metal lath welded or woven.
_ Minimum of 1/4" keying between wire and paper backing or equivalent system.

_ Furring and lath:

_ Materials   _ Sizes of furring strips or members
_ Spacing of furring strips or members   _ Alignment
_ Gauges
_ Connectors to support surfaces   _ Positions and ties
_ Plumb   _ Plain surface   _ Level members
_ Spacings of backing and frames to support finish materials

_ Metal lath and fastenings as detailed.

_ Gypsum lath and fastenings as detailed.

C. Accessories:

_ Channels as per ASTM A109 or ASTM A303.
_ Steel 26 gauge or heavier.
_ Hot-dip galvanized

_ Hot-rolled or cold-rolled channels, galvanized.

_ Cold-rolled channels:
_ flanges not less than 7/16" wide.

_ Runner channels hangers:
_ Soft annealed steel wire 8 gauge or larger.
_ As per Federal Specification QQ-W-461, galvanized.
_ Flat steel straps 3/32" x 7/8" minimum may be substituted for hanger wire.
_ Flat steel straps must be galvanized or painted with rust inhibitor.
_ Provide protective coating for ease of cleaning after plastering.

2.2 PLASTER

A. Cement Plaster:

_ Cement as per ASTM C150, Type I.
_ Lime, dry hydrated as per ASTM C206.

B. Sand as Per ASTM C144, clean and well graded from coarse to fine.

C. Water shall be clean and free from acid, alkali, and organic materials that would harm the work.

PART 3 -- CONSTRUCTION AND INSTALLATION

3.1 WORK CONDITIONS AND COORDINATION

A. Examine the job site, and correct any conditions that might interfere with speedy and acceptable work.

B. Coordinate with all other related or adjacent work such as utilities, fixtures, equipment, and finishes.

3.2 INSTALLATION -- GENERAL

*Note to specifier: Choose metal lath as appropriate to local custom or when accommodating complex, highly detailed finish plastering.

A. Install as per manufacturer's instructions, trade association standards, and governing building code.

B. If there is a conflict between instructions, standards, code, etc., install as instructed by the Architect.

3.3 INSTALLATION -- METAL LATH / GYPSUM LATH

*Note to specifier: Choose metal or gypsum board lath and edit these installation instructions accordingly.

A. Install metal lath as detailed and as instructed by the manufacturer:

_ Stagger butt joints at wall and ceiling junctures.
_ Nest ribbed lath sections together at joints.
_ Provide butt joints only where supported.
_ Use foil covered lath; face foil toward framing.

B. Install grounds and screeds as detailed and as instructed by the manufacturer:

_ Level and aligned.
_ Match to specified thickness of plaster.
_ Install in maximum lengths.
_ Do minimal, carefully done splices.
_ Install wood grounds where required.
_ Align top and bottom grounds vertically in same plane.
_ Size grounds to match combined thickness of lath plus plaster.

C. Install as much reinforcement and as many control joints needed to assure trouble-free surfaces.

_ Install diagonal lath reinforcing at stress points:

_ Window openings _ Door openings
_ Unframed openings

_ Provide relief joints at edges of small and large openings.
_ Install separation joints where work adjoins other materials and finishes.
_ Provide metal corners or edge strips to protect all exposed edges, such as at plaster corners.

_ Provide and securely anchor frames for:

_ Access panels   _ Large openings

_ Tightly install and secure lath accessories:

_ Trim _ Casings _ Corner beads _ Control joints
_ Vent screeds _ Channels _ Runners _ Stiffeners
_ Anchors _ Ties

D. Securely nail corner beads:

_ With required type and size nails.
_ Starting 2 inches from each end.
_ Spaced and staggered as required by applicable trade standard.
_ Set corner beads exactly, to assure equal plaster thickness at both sides.

E. Recheck frames and backing for wall-mounted equipment, fixtures, and specialties.

F. Gypsum lath:

_ Install gypsum lath:

_ As per manufacturer's instructions.
_ Dry
_ Staggered joints; absolutely no continuous joints from panel to panel.

*Note to specifier: Choose metal lath as appropriate to local custom or when installing reasonably simple, economical plaster.

3.4 APPLICATION AND INSTALLATION -- PLASTER

A. Examine the job site, and correct any conditions that might interfere with speedy and acceptable work.

B. Coordinate with all other related and adjacent work such as utilities, fixtures, equipment, and finishes.

_ Provide and maintain work space adequate for good work:

_ Closed in, not exposed to weather. _ Clean _ Dry
_ Protected from moisture.
_ Well-ventilated _ Well-lighted
_ Heated to required working temperature.

_ Completely protect all adjacent materials and surfaces from plaster.
_ Thoroughly cover flooring to prevent stains.
_ Protect and plug rough plumbing and drains.
_ Protect and plug open piping.
_ Protect and plug open electrical conduit.
_ Electrical outlets and switch boxes are:

_ Complete _ Positioned _ Anchored
_ Not too far in or out relative to what will be the finish wall surface.

_ Prepare concrete or masonry surfaces to be plastered:

_ Wire brush _ Abrade _ Clean _ Moisten
_ Add bonding materials or coatings.

C. Perform mixing, plastering, and plaster curing according to manufacturer's instructions.

_ Scratch coat:

_ Apply to form good keys, embedding and filling all lath openings.
_ Score to receive the brown coat.

_ Brown coat:

_ Allow 48 hours to pass after applying the scratch coat before applying the brown coat.
_ Apply brown coat to align with grounds, to a true surface
_ Provide rough surface to assure good bond of finish coat.

_ Finish coat:

_ Allow seven days to pass after applying the brown coat before applying the finish coat.
_ Apply finish coat in texture and pattern as directed and approved by the Architect.
_ Do not over-sand the mix.

D. Application:

_ Strictly observe required application sequence and curing time.
_ Verify that smoothly troweled plaster looks even and true from side views.
_ Keep textured plaster consistent across surface, and match it with sample.
_ Keep colored plaster consistent across surface, and match it with sample.

E. Tolerance for smoothness: True and even within 1/8" in 10'.

F. Cleaning during plastering:

_ Wipe metal accessories clean after application of each coat.
_ Immediately wipe clean, plaster spills on metal accessories and adjacent materials.
_ Do not track materials to other floors.

3.5 PLASTER CLEANUP AND REPAIR

A. In addition to other requirements for cleaning:

_ Inspect adjacent surfaces during and at completion of the work.
_ Remove all traces of spilled and splashed plaster.

B. Promptly remove all plastering debris from site:

_ Do not permit long-term debris storage.
_ Do not bury debris.
_ Do not store plastering debris at scrap piles.

C. Replace or repair non-conforming work before starting other room finishes.

_ Recheck work for necessary repairs before beginning painting or other added work.
_ Redo defective repairs at the direction of the Architect.
_ Make final repair work undetectable.

END OF SECTION

Notes:

DIVISION 9
FINISHES

VENEER PLASTER
09215

PART 1 -- GENERAL

*Note to specifier: Include generic introductory PART 1 text from the beginning of Division 9.

PART 2 -- MATERIALS

2.1 BASE, PLASTER AND ACCESSORIES:

A. Provide gypsum veneer plaster system where shown on the Drawings and as specified herein.

B. Comply with ASTM C754, ASTM C844, ASTM C843.

C. Provide plaster and related products manufactured by:

D. Gypsum wallboard base and related materials:

_ Wallboard shall be of types and thicknesses shown on the Drawings.
_ Wallboard shall be a minimum of 1/2" thick.
_ Provide joint tape as recommended by the manufacturer of the gypsum wallboard system.
_ Provide fasteners as recommended by gypsum wallboard manufacturer.
_ Use metal corner bead and other accessories as recommended by the wallboard manufacturer.

PART 3 -- CONSTRUCTION AND INSTALLATION

3.1 APPLICATION

A. Apply as per manufacturer's instructions

*Note to specifier: If using a two-coat system describe from manufacturer's specifications:

_ First coat: 1/16" thick.
_ Second coat: 1/16" thick.

_ One-coat system *(product name and manufacturer) 3/32" thick

B. Machine mix plaster and trowel on to provide dense finish surface.

  _ Texture or special surfaces will be as per Finish Schedule and as directed by the Architect.
  _ Provide plaster finish behind equipment, casework and similar removeable items.
  _ Finish may be omitted from permanently concealed surfaces.

C. Tolerance for smoothness: true and even within 1/8" in 10'.

3.5 CLEANING AND REPAIR

A. Cleaning during plastering:

_ Wipe metal accessories clean after application of each coat.
_ Immediately wipe clean, plaster spills on metal accessaries and adjacent materials.
_ Do not track materials to other floors.

B. In addition to general requirements for maintaining clean working conditions:

_ Inspect adjacent surfaces during and at the completion of the work.
_ Remove all traces of spilled and splashed plaster.
_ Promptly remove all plastering debris from site.

C. Replace or repair non-conforming work before starting other room finishes.

_ Recheck work for necessary repairs before beginning painting or other added work.
_ Redo defective repairs at the direction of the Architect.
_ Make final repair work undetectable.

END OF SECTION

Notes:

DIVISION 9
FINISHES

GYPSUM WALLBOARD
09250

PART 1 -- GENERAL

*Note to specifier: Include generic introductory PART 1 text from the beginning of Division 9.

PART 2 -- MATERIALS

2.1 GYPSUM WALLBOARD

A. Gypsum wallboard shall be manufactured by:

*Note to specifier: As appropriate, add material or product type, trade name, grade, size, manufacturer, and manufacturer representative's location.

_ Provide boards in 8 foot or other lengths for a minimum of joints.

B. Gypsum wallboard shall be as per Federal Specification SS-L-30D, in 48" widths.

C. Gypsum wallboard sheathing as per Federal Specification SS-L-30D, Type II, Grade W, Class 2.

D. Use types and thicknesses specified below except where shown otherwise in the Drawings.

_ Standard wallboard: Type III, Grade R, Class 1, 5/8" thick.
_ Fire-retardant wallboard: Type III, Grade X, Class 1, 5/8" thick.
_ Water-resistant wallboard: Type VII, Grade W or X as required, Class 2, 5/8" thick.
_ Foil-backed wallboard: Provide where shown on Drawings.
_ Shaft walls, as indicated on Drawings.
_ Wallboard designed for shafts of required fire-resistance.
_ As per Federal Specification SS-L-30D, Type IV, Grade R or X, Class 1.

_ Provide seals for sound and thermal insulation at:
_ Floor plates   _ Top plates
_ Connection to adjacent walls/pilasters/columns
_ All cutouts

2.2 METAL TRIM AND ACCESSORIES

A. Metal Trim:

_ Zinc-coated steel 26 gauge minimum, as per Federal Specification QQ-S-775, Class d or e.

B. Casing beads:

_ Channel-shapes with exposed wing, and concealed wing not less than 7/8" wide.

C. Corner beads:

_ Angle shapes with wings not less than 7/8" wide:
_ Perforated for nailing and joint treatment.
_ Or use paper/metal combination bead suitable for joint treatment.

D. Edge beads at ceiling perimeter:

_ Angle shapes with wings 3/4" wide minimum.
_ Concealed wing perforated for nailing, exposed wing edge folded flat.

2.3 JOINTING

A. Jointing system with reinforcing tape and compound:

_ As supplied or recommended by the gypsum wallboard manufacturer.

2.4 FASTENINGS

A. For gypsum wallboard attached to metal framing and channels:

_ Flat-head screws, 1" long minimum.
_ Self-tapping threads and self-drilling points.
_ Specifically designed for use with power-driven tools.

B. For gypsum wallboard attached to wood:

_ 1-1/4" type W bugle-head screws.
_ Alternate: Annular ring nails complying with ASTM C514.
_ Nail sizes as required by governing building code.

2.5 ACCESS DOORS AND PANELS

A. Access doors and panels in plastered walls and ceilings:

_ Install where shown on Drawings and approved by the Architect.
_ Install as necessary for maintenance access to mechanical and electrical work.

B. Types of access panels and doors:

_ 24" x 24" metal access doors, concealed hinges, Allen key lock.
_ At fire-rated surface, access doors shall have the same fire rating as surface.
_ At tile surfaces and toilet rooms, use stainless steel doors and frames with satin finish.

C. Other installations:

_ Prime-coat steel access doors and frames for finish painting at the site.

PART 3 -- CONSTRUCTION AND INSTALLATION

3.1 PREPARATION

A. Preparation and coordination:

_ Install blocking and backups to support all edges of wallboard.
_ Verify that wood framing to receive wallboard is dry and not subject to shrinkage.

B. Work conditions:

_ Keep wallboard materials dry and protected from moisture.
_ Store wallboard materials so they are protected from damage to surfaces and edges.
_ Do not overload floor construction with stored wallboard.
_ Estimate stored material weights, and assure that dead load of stored material is not excessive.
_ Maintain interior work environment:
_ Closed in, not exposed to weather _ Clean _ Dry
_ Protected from moisture
_ Well-ventilated _ Well-lighted
_ Comfortable in temperature

_ Keep work of trades such as conduit, pipe, and ducts clear of the inside faces of wall panels.

3.2 INSTALLATION

A. Install as per manufacturer's instructions, trade association standards, and governing building code.

B. If there is a conflict between instructions, standards, code, etc., install as instructed by the Architect.

C. For walls and ceilings:

_ Hold wallboard 3/8 inch to 1/2 inch up from floor.
_ Install wall panels horizontally unless otherwise required.
_ Stagger panel joints vertically.
_ Stagger panel joints back-to-back if using double-layered panels.
_ Keep long joints of ceiling joints straight and aligned.
_ Stagger short joints of ceiling panels at half the long dimensions of panels.
_ Keep joints to a minimum.
_ Align door jambs and vertical joints.
_ Keep piecing and use of odd sizes to an absolute minimum.
_ Remove damaged sheets as scrap.
_ Use moisture-resistant wallboard in damp environments.
_ Seal edges and cuts of wallboard in damp environments.
_ Install metal corners and other protective strips where finish wallboard edges might be damaged.
_ Never force wallboards into place.
_ Install gypsum wallboard at right angles to furring or studs.
_ All end joints over framing or furring members.
_ Install wallboard to ceilings with long dimension of board at right angles to joists.

D. Nailing and screw attachment:

_ Attach screws with clutch-controlled power screwdrivers.
_ Attach screws at 12" o.c. at ceilings and 16" o.c. at walls.
_ Where wall framing members are 24" apart, space screws 12" on center.
_ Start nailing from centers of panels, and proceed outward to edges.
_ Avoid paper damage in nailing.
_ Where paper damage occurs, add another nail within about 2 inches.
_ Do not position conduit and piping where it can be damaged by nailing.
_ Do not proceed with nailing into wood framing that has over 19% of moisture content.

E. Sealing:

_ Thoroughly seal penetrations in fire-rated walls.
_ Box in recesses in fire-rated walls.
_ Cut cutouts for electrical outlets, switch boxes, pipe, etc., tightly to size.

F. Taping and Spackling:

_ Follow applicable trade standards and manufacturer's instructions throughout.
_ Keep temperature above specified minimum (usually 55 degrees).
_ Do not allow any bumps, bubbles, or dimples in taping and spackling.
_ Allow 24 hours between spackling coats.
_ Sand and spackle wallboard as required, as base for final specified texture or finish.
_ Do not track gypsum and spackle dust to clean areas.
_ Final spackle coat:
_ Sand smoothly
_ Feather outward from 12 inches to 16 inches, each side of joint.

G. Joint Treatment:

_ Gypsum wallboard must fit completely snugly against supporting framework.
_ Joint work shall be at a minimum of 55 degrees F. for 24 hours prior to work.
_ Joint and finishing compounds must dry 24 hours before starting finishing.
_ Apply joint treatment and finishing compound by machine or hand tool.
_ Allow minimum drying time of 24 hours between coats.
_ Allow additional drying time in humid or poorly ventilated areas.
_ Embedding compounds:
_ Apply at wallboard joints and fastener heads in thin uniform layer.
_ Spread compound not less than 3" wide at joints.

_ Finishing compounds:
_ Apply to joints and fastener heads when compound is dry and has been sanded.

_ Sandpaper between coats, and do final sandpaper work to eliminate all ridges and high points.
_ Feather finishing compound to not less than 12" wide.
_ When thoroughly dry, sandpaper to a uniform smooth surface without damaging wallboard.

3.3 ACCESS DOORS:

A. Install access doors in coordination with other work and as approved by the Architect.

_ Anchor access doors firmly into position.
_ Install access doors to be completely straight, flush, and aligned with finished surface.

3.4 METAL TRIM

A. Provide all metal trim as required to complete the work.

_ Securely nail corner beads:

_ With required type and size nails.
_ Starting 2 inches from each end.
_ Space and stagger as required by wallboard system manufacturer.

3.5 CLEANING AND REPAIR

A. Cleaning:

_ Don't allow tracking of gypsum and finishing compounds onto floor surfaces.
_ Don't allow gypsum and compound dust to accumulate at edges of tarps or protective plastic.
_ At completion of each segment of work in a room, clean thoroughly and remove all debris.
_ Frequently remove all debris from site.
_ Do not allow gypsum dust to blow at site storage of scraps.
_ Make a final check to determine that there are no penetrations through fire-rated walls.

B. Repair:

_ Recheck work for necessary repairs that may be required before painting or other added work.
_ Complete repairs so they are undetectable.

END OF SECTION

Notes:

_____

_____

_____

_____

DIVISION 9
FINISHES

SAMPLE CERAMIC TILE CONSTRUCTION

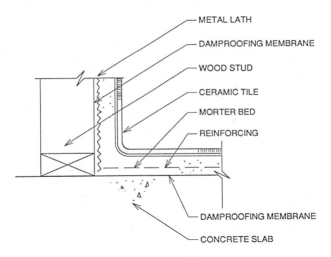

TILE FLOOR @ BASE

DIVISION 9
FINISHES

CERAMIC TILE
09300

QUARRY TILE
09330

PART 1 -- GENERAL

*Note to specifier: Include generic introductory PART 1 text from the beginning of Division 9.

PART 2 -- MATERIALS

2.1 TILE AND ACCESSORIES

A. Tile:

_ Shall comply with Tile Council of America Specification 137.1.
_ Colors, textures, and patterns will be as selected by the Architect from manufacturer's samples.
_ Delivered tile shall match samples approved by the Architect.

B. Wall tile shall be manufactured by:

*Note to specifier: List tile type, color, texture, unit sizes, and manufacturer.

*Note to specifier: Note special tiles such as non-conductive or non slip. Refer to manufacturer's specifications.

C. Floor tile shall be manufactured by:

*Note to specifier: List tile type, color, texture, unit sizes, and manufacturer.

_ Floor tile shall have coefficient of friction not less than 0.50 as per ASTM F489, ASTM F609.
_ Floor tile as per National Bureau of Standards Technical Note 895.

D. Tile for countertops shall be manufactured by:

*Note to specifier: List tile type, color, texture, unit sizes, and manufacturer.

E. Quarry tile conforming to ANSI A137.1.

*Note to specifier: If quarry tile work is extensive, prepare a separate specification section and rely on manufacturers specifications. Typical quarry tile is 6" by 6" by 1/2", unglazed slip-resistant square edged.

F. Provide all accessory shapes to complete the work as shown in Drawings and specified herein.

2.2 SETTING MATERIALS

A. Comply with Tile Council of America "Handbook for Ceramic Tile Installation."

B. Latex-portland cement mortar as per ANSI A118.4.

C. Organic adhesive as per ANSI A136.1.

_ Type I where subject to extended water exposure.
_ Type II at all other locations.

2.3 GROUT

A. Comply with Tile Council of America "Handbook for Ceramic Tile Installation."

B. Colors will be selected by the Architect from standard colors available from approved manufacturers.

C. Portland cement grout:

_ Use mixture of portland cement and other materials manufactured for this purpose.
_ Grout must comply with tile manufacturer's instructions.

D. Obtain Architect's approval of any proposed material.

2.4 OTHER MATERIALS

A. Provide all other materials, required for a complete, proper installation.

B. Adhesive, sealant, and grout as per applicable trade standards and tile manufacturer's instructions:

_ Delivered in new unopened containers.
_ With correct color additives.

C. Provide non-corrosive lath:

_ Zinc-coated _ Lapped _ Tied with zinc-coated wires

D. Install waterproofing and backing that will absolutely block water leakage.

_ All waterproofing and backing as per manufacturer's instructions in:

_ Primer _ Felt _ Waterproof membrane _ Sealant
_ Waterproof sheet metal work
_ Underlayment

PART 3 -- CONSTRUCTION AND INSTALLATION

3.1 PREPARATION

A. Keep work surfaces and working environment:

_ Clean _ Dry _ Well lighted _ Well ventilated
_ Free of airborne construction dust.
_ At comfortable working temperature, minimum 60 degrees.

B. For accessories and other work:

_ Provide supports for fixtures and related construction:
_ Wall blocking _ Backing _ Inserts _ Anchors

_ Pre-mark and double-check locations for accessories to be installed.
_ Set accessories in place before beginning tile work.

_ Put in place and properly position, work of related trades:
_ Electrical outlet and switch boxes
_ Stub ups _ Drains _ Wall plumbing

C. Install all support framing, furring, and backing:

_ Plumb _ Square _ Aligned
_ Well secured so surfaces will not move or deflect.

D. Prepare floors for tiling so the finish floor will be either perfectly level or slope properly to drains.

E. Work preparation:

_ Install waterproofing and backing that will absolutely block water leakage.

_ Install control joints and edge strips:
_ Securely fastened.
_ Sized and shaped exactly to fit finish tile work.

_ Set layout start points to achieve tile patterning that is:
_ Symmetrical and complete.
_ Planned for minimal and balanced tile cutting.

_ Include trim, edge, and base shapes with materials that:
_ Match other tiles in all regards.
_ Are free of irregularities.

3.2 INSTALLATION

A. Work standards and conditions:

_ Comply with Tile Council of America "Handbook for Ceramic Tile Installation."
_ Comply with ANSI A108.1, ANSI A108.2.
_ Comply with manufacturer's instructions.
_ Work temperature must be as per instructions of materials manufacturers.
_ Tile over floor membrane may not be installed until membrane is tested and accepted.

B. Limits of tile application:

_ Extend tile into recesses and under and behind future equipment or fixtures.
_ Tile must be installed as a complete, uninterrupted covering.
_ Edges and corners must be terminated neatly without disruption of pattern or joint alignment.
_ Terminate tile neatly at obstructions or penetrations of other work.

C. Joint pattern:

_ Lay tile in standard grid unless shown otherwise on Drawings or directed by the Architect.
_ Align joints of adjoining same size tiles on floor, base, walls, and trim.
_ In tile layout, center tile fields both directions on each floor or wall area.
_ Adjust layout and pattern to minimize tile cutting.
_ Joint widths must be consistent and uniform.

D. Provide expansion and control joints:

_ Where shown on Drawings.
_ As instructed by the "Handbook for Ceramic Tile Installation" of the Tile Council of America.

E. Standards of work quality:

_ Fit special border tiles squarely without cuts.

_ Perfectly match tile pieces with other tile work:

_ Trim   _ Tile accessories   _ Bases

_ Prepare tile according to manufacturer's instructions:

_ Presoak   _ Dry surfaces before application   _ Clean

_ Plan and install correct patterning:

_ Symmetrical   _ Complete   _ Square to floor or wall

_ Apply tile surface smoothly and free of:
_ Irregularities   _ Humps   _ Dips

_ Install tile joints:

_ Straight   _ Level horizontally
_ Aligned and exact vertically   _ Uniform in size

_ With extra care at difficult areas:

_ Corners   _ Fixture locations
_ Around wall openings and recesses
_ At penetrations such as floor drains   _ Door trim

_ Make tile cuts:

_ Minimal   _ Uniform   _ Not smaller than half a tile.

_ Complete grouted or thin-set adhesion so no tiles can be pulled loose.
_ Do not allow tile at door thresholds to interfere with closure.
_ Do not use broken or cracked scrap tiles.

3.3 PROTECTION, CLEANING, AND REPAIR

A. Completely protect finished tile, and allow no damage to the work.

B. Cleaning:

_ Wash tile surfaces with clean water before and after cleaning.
_ Use cleaning solutions and materials as per manufacturer's instructions.
_ Do not use acidic cleaners near finish metal or other vulnerable surfaces.
_ Remove excess corrosive cleaning solutions from site; do not empty into building drains.

C. Repair and replace defective work:

_ Reject tiles and replace if:

_ Chipped _ Scratched _ Crazed _ Popped up
_ Loose   _ Stained   _ Misaligned

_ Repair or replace all defective and non-conforming work as directed by the Architect.
_ Make repairs undetectable.

END OF SECTION

Notes:

___
___
___
___
___
___
___
___
___
___
___
___
___
___

DIVISION 9

FINISHES

TERRAZZO
09400

PART 1 -- GENERAL

*Note to specifier: Include generic introductory PART 1 text from the beginning of Division 9.

PART 2 -- MATERIALS AND ACCESSORIES

*Note to specifier: Rely on manufacturer's specifications for complete data on materials and application.

2.1 MATERIALS

A. Cementitious terrazzo:

_ Sand cushion cast-in-place terrazzo, 2-3/4" - 3" thick.
_ Bonded cast-in-place terrazzo, 1-1/2" - 2" thick.
_ Monolithic cast-in-place terrazzo, 1/2" thick.
_ Precast terrazzo for stair treads, bases, trim.

B. Thin-set terrazzo, modified cementitious or resinous matrix.

C. Accessories: Half-hard brass dividers and accessories.

_ Materials per approved samples:
  _ Aggregates _ Inserts _ Reinforcing _ Divider strips
  _ Expansion strips _ Precast materials

PART 3 -- INSTALLATION

*Note to specifier: Rely on manufacturer's specifications for complete data on preparation and application. Terrazzo application must be done only by well trained personnel with experienced subcontractors.

3.1 PREPARATION

A. Terrazzo substrate per manufacturer's instructions or applicable trade standards:

_ Prepare surface as required for terrazzo bonding:
_ Clean _ Wet prior to application

_ Position joint materials of correct height:
_ Beads _ Temporary screeds _ Expansion strips

_ Match pattern layout with intended design.
_ Line up terrazzo joints with adjacent construction.

3.2 INSTALLATION

A. Apply materials and allow them to set per manufacturer's instructions and trade standards.

B. Cure exactly as per manufacturer's instructions or applicable trade standards.

C. Properly equip for grinding operation:

_ Adequate, well-trained work crew.
_ Correct equipment.
_ Grinder dust control system.

D. Finish terrazzo surface:

_ Level, except as sloped to drain.
_ Do not allow humps or depressions.
_ Do not allow irregularities.
_ Keep smooth and true throughout.
_ Provide nonslip surfaces where required for pedestrian safety.

E. After curing and grinding, verify that finish work is:

_ Completely cleaned   _ Finish sealed
_ Protected from construction work

3.3 PROTECTION, CLEANING, AND REPAIR

A. Completely protect finished tile and allow no damage to the work.

B. Cleaning:

_ Wash surfaces with clean water before and after cleaning.
_ Use cleaning solutions and materials as per manufacturer's instructions.

C. Repair and replace defective work:

_ Repair or replace all defective and non-conforming work as directed by the Architect.
_ Make repairs undetectable.

END OF SECTION

Notes:

_____

_____

_____

_____

DIVISION 9
FINISHES

INTERIOR STONE
09600

PART 1 -- GENERAL

*Note to specifier: Include generic introductory PART 1 text from the beginning of Division 9.

A. Provide interior stonework as shown on the Drawings and specified herein.

*Note to specifier: identify areas and types of interior stone work.

_ Wall facing and trim.
_ Stone flooring.
_ Steps and platforms.
_ Lavatory counters and vanities, including splashbacks.
_ Facing for architectural woodwork.

PART 2 -- MATERIALS AND ACCESSORIES

2.1 STONE:

*Note to specifier: A few sample stone types and trade standard references are shown below. Rely on supplier's specifications for complete data on materials and application.

_ Marble: MIA Group A, polished finish.
_ Granite: ASTM C 615, architectural grade, thermal or polished finish.
_ Slate: ASTM C 629, even grain, natural finish.
_ Bluestone: Hard, fine grained, uniform color, natural finish.
_ Limestone: ASTM C 568 Indiana limestone, smooth dense surface.

PART 3 -- APPLICATION

3.1 SETTING METHODS

*Note to specifier: Chose a setting method and use manufacturer's specifications and instructions to complete this section:

A. Floors: _ Thin set, latex-modified portland cement.
_ Mortar set, portland cement.

Walls: _ Anchor set, stainless steel anchors and plaster spots.
_ Mortar set, portland cement.

3.2 PROTECTION, CLEANUP AND REPAIR

B. Completely protect finished tile and allow no damage to the work.

C. Cleanup and repair:

_ Repair or replace all defective and non-conforming work as directed by the Architect.
_ Make repairs undetectable.

END OF SECTION

DIVISION 9
FINISHES

ACOUSTICAL INSULATION
09530

PART 1 -- GENERAL

*Note to specifier: Include generic introductory PART 1 text from the beginning of Division 9.

PART 2 -- MATERIALS AND ACCESSORIES

2.1 Provide and install acoustical insulation material as detailed:

_ Types _ Thicknesses _ Backing materials _ Stored dry

_ Completely fill all batt spaces.

_ Repack any cutouts made through insulation.

_ Tightly attach edges to surfaces.

PART 1 -- GENERAL

Note to specifier: Use specifications as provided by acoustical consultant.

END OF SECTION

Notes:

_____
_____
_____
_____
_____
_____
_____
_____
_____

DIVISION 9
FINISHES

WOOD FLOORING
09550

PART 1 -- GENERAL

*Note to specifier: Include generic introductory PART 1 text from the beginning of Division 9.

1.2 SUBMITTALS

   A. Submit for approval samples, product data, warranty, maintenance data, extra stock.

PART 2 -- MATERIALS AND ACCESSORIES

2.1 WOOD FLOORING AND ACCESSORIES

A. Wood strip flooring:

   _ Select grade plain-sawn Red oak or white oak, 25/32" thick; 3-1/4" face width.
   _ Tongue and groove, end matched.
   _ "Prime" grade.

B. Nails, screws, other fastenings as per flooring manufacturer's instructions.

_ Ring-shank flooring nails must be long enough to securely attach the flooring to substrate.
_ Nails must not split the flooring.

PART 3 -- INSTALLATION

3.1 PREPARATION:

A. Protection and coordination:

_ Store wood flooring materials in dry, protected work space 72 hours prior to installation.

_ Confirm Drawings and specifications with subcontractor:

_ Materials _ Strip direction _ Patterning _ Bordering
_ Nailing, types and spacing
_ Scheduling of sanding and finish

3.2 APPLICATION:

A. Install flooring as detailed:

_ Reject warped or bent material.
_ Match with samples.

B. Install sleepers as detailed:

_ Sizes _ Spacings _ Fastening _ Ventilation

C. Nails and nailing:

_ Drive diagonally.
_ Space as required.
_ Nail at ends of each strip.
_ Pre-drill as necessary to prevent splits.
_ Nail type as instructed by manufacturer

D. Joints:

_ Construct joints within tolerances required by manufacturer.
_ Do not allow end joints to occur side by side; separate by at least two strips.
_ Construct tight joints.
_ Do not damage tongues and grooves before or during application.
_ Use small or varied strips sparingly and never near one another.
_ Provide expansion joint space at all walls (1/2 inch min.).
_ Fasten baseboards at walls only, not at floors, to cover expansion joint space.

3.3 CLEANUP, FINISH, PROTECTION AND REPAIR

A. Cleaning:

_ Keep work area thoroughly clean:

_ Clear all nails away.
_ Remove all scrap.

_ Sand promptly after installation and cleaning:

_ Schedule to avoid contaminating other work.
_ Consistently smooth.
_ Without lumps.
_ Without depressions.
_ Without burns.

_ Thoroughly vacuum away all sanding dust.

B. Final finish:

_ Apply final finish as per manufacturer's instructions:

_ Filler _ Stain _ Wax and buffing

_ Do final finish immediately after sanding and cleaning.

C. Protection:

_ Cover and protect floor surfaces from:
_ Construction equipment.
_ Materials storage and movement.
_ Foot traffic.
_ Paint and other spills/droppings.
_ Temperature extremes.
_ Weather intrusion through open doors and/or windows.
_ Any other sources of moisture.

D. Repair:

_ Repair or replace defective work.
_ Make repairs undetectable.

END OF SECTION

Notes:

DIVISION 9
FINISHES

WOOD FLOORING -- PARQUET AND BLOCK
09570

PART 1 -- GENERAL

*Note to specifier: Include generic introductory PART 1 text from the beginning of Division 9.

PART 2 -- MATERIALS AND ACCESSORIES

*Note to specifier: Choose parquet floor and delete those not to be used

2.1 FLOORING AND ACCESSORIES

A. Flooring:

_ Plastic-impregnated parquet flooring:

_ Solid oak slat parquet flooring impregnated with acrylic plastic, 5/16" thick.
_ Tongue and groove edges, factory finish.

_ Hardwood parquet flooring:

_ Prime grade quarter-sawn white oak.
_ Solid stock strips pre-assembled into 9" square units, 5/16" thick.
_ Tongue and groove edges, factory finished

_ Teak parquet flooring:

_ Prime grade teak.
_ Solid stock strips pre-glued into 12" x 12" square units, 5/16" thick.
_ Square edges, factory finished.

B. Adhesives as per manufacturer's instructions:

_ Type _ Manufacture _ New
_ Unopened containers
_ Special application as per manufacturer's instructions.
_ Keep work space well ventilated.

PART 3 -- CONSTRUCTION AND INSTALLATION

3.1 APPLICATION

A. Parquet flooring strictly as per manufacturer's instructions.

_ Take special care in storage and handling to avoid damage to parquet floor units.

B. Block Flooring as per manufacturer's instructions:

_ Apply fully bedded.
_ Do not butter edges.
_ Provide prefinished blocks free of raised edges.

3.3 CLEANUP, FINISH, PROTECTION AND REPAIR

A. Cleaning.

_ Keep work area thoroughly clean:

_ Clear all nails away.
_ Remove all scrap.

_ Sand promptly after installation and cleaning:

_ Schedule to avoid contaminating other work.
_ Consistently smooth.
_ Without lumps.
_ Without depressions.
_ Without burns.

_ Thoroughly vacuum away all sanding dust.

B. Final finish.

_ Apply final finish as per manufacturer's instructions:

_ Filler   _ Stain   _ Wax and buffing

_ Do final finish immediately after sanding and cleaning.

C. Protection:

_ Cover and protect floor surfaces from:

_ Construction equipment.
_ Materials storage and movement.
_ Foot traffic.
_ Paint and other spills/droppings.
_ Temperature extremes.
_ Weather intrusion through open doors and/or windows.
_ Any other sources of moisture.

D. Repair:

_ Repair or replace defective work.
_ Make repairs undetectable.

END OF SECTION

DIVISION 9
FINISHES

ASPHALT, VINYL, AND RESILIENT TILE FLOORING
09660

*Note to specifier: Except for the specific flooring material, specifications for these floors are virtually identical. Refer to text below.

PART 1 -- GENERAL

*Note to specifier: Include generic introductory PART 1 text from the beginning of Division 9.

PART 2 -- MATERIALS

2.1 FLOORING AND ACCESSORIES

A. (Tile type) shall be: as manufactured by:

*Note to specifier: Name the selected flooring type -- asphalt, vinyl, or generic "resilient tile."

_ Flooring shall match samples selected by the Architect from approved manufacturer.
_ Equal products of other manufacturers may be submitted for approval by the Architect.

B. RESILIENT TILE FLOORING SCHEDULE

Room	Manufacturer	Product	Color	Pattern

C. RUBBER BASE SCHEDULE

Room	Manufacturer	Product	Color	Pattern

D. Where shown on the Drawings, provide top-set base type manufactured by:

_ Material shall match samples selected by the Architect from approved manufacturer.

E. Adhesives:

_ Waterproof stabilized adhesive as recommended by the flooring manufacturer.
_ Do not use asphalt emulsions or other non-waterproof adhesives.

F. Maintenance supplies:

_ Provide 5% of the total number of tiles to the Owner, well wrapped and clearly labeled.

PART 3 -- CONSTRUCTION AND INSTALLATION

3.1 PREPARATION

A. Subfloor or substrate:

_ Smooth _ At required finish elevation
_ No more than 1/8" in 10'-0" deviation from level or slopes shown on Drawings.
_ Sweep or vacuum clean and inspect for smoothness and any needed repairs.

B. Apply concrete slab primer:

_ Concrete slab primer must be non-staining and recommended by the flooring manufacturer.
_ Apply primer as per manufacturer's instructions.

3.2 INSTALLATION

A. Install resilient tiles as per manufacturer's instructions and as approved by the Architect.

B. Install flooring after:

_ Other finish work such as painting is completed.
_ Building's heating system is operational.
_ Moisture in surfaces and interior climate is within manufacturer's limits including:
_ Concrete slab moisture _ Building air temperature
_ Relative humidity

C. Install:

_ Thoroughly adhere base materials; do not spot glue.
_ Use materials from cartons in the same sequence manufactured and packaged.

- Lay units from center marks oriented to principal walls
- Lay units square relative to principal walls.
- Lay in pattern with grain direction as directed by the Architect.
- Butt tile units tightly against vertical surfaces, nosings, etc.
- Scribe as necessary to fit around objects.
- Scribed joints must be as neat and square manufactured edges.
- Units at opposite ends of each room will be equally wide.
- Avoid cut widths less than 3" at edges of rooms.
- Chalk mark locations of openings to be cut for pipe, etc. as flooring is laid.
- Place edge strips butted tightly to tiles and fastened with adhesive at unprotected edges.
- Do not allow scrap for base work; use maximum piece lengths.

- Place tiles neatly:

- Tightly aligned _ Even _ In straight parallel lines

- Extend units into recesses:

- Toe spaces _ Door reveals

- In concealed spaces such as behind fixtures

3.3. CLEANING AND FINISHING

A. Only use cleaner recommended by the flooring manufacturer.

- Remove excess adhesive and other marks or stains from finish flooring.
- Remove all stains and excess adhesive immediately after installation.

B. Leave factory finish unless otherwise required.

- Do new finishing such as waxing strictly according to manufacturer's instructions.

3.4 REPAIR

A. Repair or replace defective work:

- Chipped _ Scratched _ Marred _ Wrinkled _ Cracked
- Blistered
- Edges or corners are raised _ Joints are not completely tight
- Work not level

- No gaps at:

- Walls _ Jambs _ Trim

- Verify level joining at flush floor electrical cover plates, cleanouts, etc.
- Make repairs undetectable.
- Securely protect floor from damage by traffic or further construction work.

END OF SECTION

DIVISION 9
FINISHES

RESILIENT SHEET FLOORING
09665

PART 1 -- GENERAL

*Note to specifier: Include generic introductory PART 1 text from the beginning of Division 9.

PART 2 -- MATERIALS

*Note to specifier: Adapt text from the Resilient Tile section to complete this section on sheet flooring.

2.1 SHEET FLOORING AND ACCESSORIES

A. Resilient sheet flooring shall be:

as manufactured by:

B. VINYL SHEET FLOORING SCHEDULE:

Room   Manufacturer   Product   Color   Pattern

PART 3 -- CONSTRUCTION AND INSTALLATION

3.1 APPLICATION

A. Install resilient sheet flooring so that:

_ All portions are laid in one uniform direction.
_ Color and pattern and matched throughout installation.
_ Seams are minimal.
_ Seam joints are:
_ Cleanly cut _ Straight _ Matched _ Aligned
_ Level, without humps or depressions

B. Install flooring as per manufacturer's instructions:

_ Unroll rolled material 24 hours ahead of installation, if required by
    manufacturer's instructions.
_ Start compression rolling over sheet flooring in middle; move outward to
    press out all bubbles.

3.2 CLEANING AND FINISHING

A. Only use cleaner recommended by the flooring manufacturer.

_ Remove excess adhesive and other marks or stains from finish flooring.
_ Remove all stains and excess adhesive immediately after installation.

B. Leave factory finish unless otherwise required.

_ Do new finishing such as waxing strictly according to manufacturer's instructions.

3.3 REPAIR

A. Repair or replace defective work:

_ Chipped _ Scratched _ Marred _ Wrinkled _ Cracked
_ Blistered
_ Edges or corners are raised _ Joints are not completely tight
_ Work not level
_ No gaps at:
_ Walls _ Jambs _ Trim

_ Verify level joining at flush floor electrical cover plates, cleanouts, etc.
_ Make repairs undetectable.
_ Securely protect floor from damage by traffic or further construction work.

END OF SECTION

Notes:

DIVISION 9
FINISHES

CARPET
09680

PART 1 -- GENERAL

*Note to specifier: Include generic introductory PART 1 text from the beginning of Division 9.

PART 2 -- MATERIALS AND ACCESSORIES

2.1 CARPET

A. Provide and install (material) carpet manufactured by:

*Note to specifier: As appropriate, add trade name and manufacturer representative's location.

B. CARPET SCHEDULE:

Room	Manufacturer	Product	Color	Pattern

C. Carpet shall have:

*Note to specifier: Fill in these attributes from the selected carpet manufacturer's specifications, or incorporate the manufacturer's specifications by reference.

_ Surface:
_ Flame spread:
_ Smoke developed:
_ Weave:
_ Gauge:
_ Stitches:
_ Pile height:
_ Face yarn:
_ Face yarn weight:
_ Total weight:
_ Backing material:
_ Dye method:
_ Static parameter:
_ Installation method:

D. Provide all related materials and accessories required for the work.

E. Adhesives:

_ Provide adhesive as recommended by the carpet manufacturer.
_ Provide seam adhesive as recommended for use by the carpet manufacturer.
_ At intersections of carpet and other flooring provide carpet strips as selected by the Architect.

PART 3 -- CONSTRUCTION AND INSTALLATION

3.1 PREPARATION

A. Order carpet in ample time for scheduled installation.

B. Verify that materials are delivered undamaged; store:

_ Protected from weather and moisture.
_ Protected from construction dirt and damage.

C. Make subfloor ready for finish flooring:

_ Eliminate irregularities and high spots, fill low spots up to an flush even surface.
_ Remove grease, paint, varnish, and any other material that might interfere with the adhesive.
_ Nail well, and secure against moving and squeaking.
_ Patch, repair, and sand smooth.
_ Tighten and seal at joints.
_ Clean substructure or underlayment.
_ Make substructure level.

D. Make concrete slab ready for finish flooring:

_ Level, without humps or dips.
_ Dry (test moisture content if required).
_ Trowel smooth.
_ Patch and repair as required for smooth surface.
_ Seal smoothly at joints.
_ Non-powdery.
_ Clean.

E. Prepare joints and coordinate with carpet installation:

_ Thresholds   _ Carpet strips
_ Borders with other flooring materials
_ Trench duct cover plates   _ Floor access panels
_ Cuts at openings and edges such as balcony or stair.
_ Cuts at: _ Stub ups   _ Steps   _ Platforms   _ Curbs

F. Coordinate carpet installation with other trades:

_ Baseboard installation.
_ Floor-mounted fixtures.
_ Floor-mounted cabinets and furnishings.

3.2 INSTALLATION

A. Install carpet so that:

_ All portions are laid in the same direction unless specifically directed by the Architect.
_ There are no fill strips less than 6" wide,

B. Place seams where shown on accepted Shop Drawings or as directed by the Architect.

_ Seams are minimal.
_ There are no seams in heavy traffic areas.
_ There are no seams in high visibility areas.
_ There are no cross seams.
_ Cuts for seams are made only on the weave line.
_ Fabricate seams with compression method.

_ Use butt joints, thoroughly bedded and sealed.
_ Cuts are not visible at lines of walls, door jambs, floor-mounted cabinets, etc.
_ Seam joints are:
_ Straight _ Matched _ Aligned
_ Level, without humps or depressions

C. Carpet padding:

_ Apply carpet pad per manufacturer's instructions:
_ Protect from moisture before installation.
_ Install perfectly dry.
_ Adhere thoroughly.
_ Adhere with materials required by manufacturer.

_ Prepare carpet and pad prior to installation:
_ Unroll 24 hours in advance.
_ Check for defects and moisture.

D. Wall carpet:

_ Install wall carpeting as directed by manufacturer.
_ Terminate exposed edges of wall carpet with trim as directed by the Architect.

3.3 CLEANING AND PROTECTION

A. Cleaning:

_ Clean scraps, threads, and dust as work proceeds.
_ Thoroughly clean carpet and adjacent surfaces upon completion of installation.

B. Protection:

_ Do not expose to weather or moisture.
_ Provide heavy duty non-staining paper, plastic, or board walkways as directed by the architect.
_ Allow no damage to the carpet from traffic, spills, or other work.
_ Damaged work will be replaced by the Contractor at no cost to the Owner.
_ Replaced or repaired carpet will be undetectable.

3.4 REPAIR AND SURPLUS MATERIAL

A. Surplus:

_ Save large scraps for owner's maintenance.
_ Return for credit or otherwise use for benefit of owner, any large amount.
_ Wrap selected scraps in burlap and deliver to Owner.

B. Repair:

_ Repair or replace all defective and non-conforming work.
_ Make repairs so they are undetectable.

END OF SECTION

DIVISION 9
FINISHES

PAINTING
09900

PART 1 -- GENERAL

*Note to specifier: Include generic introductory PART 1 text from the beginning of Division 9.

1.3 SUBMITTALS

A. Provide manufacturer's specifications and other data to prove compliance with specified requirements.

B. Paint and color samples:

_ Following selection of colors by the Architect, submit samples for the Architect's review.
_ Provide three samples of each color and gloss for each material.
_ Samples shall be on the material the finish is specified to be applied.
_ Samples shall be approximately 8" x 10" in size.
_ Provide samples of paint on actual surfaces to be painted as directed by the Architect.
_ Revise and resubmit samples as requested until colors, gloss, etc. are approved by the Architect.
_ Do not start finish painting until samples are approved and available at job site.

1.4 JOB CONDITIONS

A. Unless specifically allowed by paint manufacturer, do not apply paint:

_ When temperature of work surfaces and air temperature
_ When weather is inclement with snow, rain, or mist.
_ When relative humidity exceeds 85%.
_ To damp or wet surfaces.

PART 2 -- MATERIALS

2.1 PAINT AND RELATED MATERIALS

A. Provide all materials and tools required for the work.

B. Paints shall be as per paint schedule.

EXTERIOR PAINTING SCHEDULE

	First Coat	Second Coat	Third Coat

EXTERIOR WALLS:

Manufacturer

Type

Color

EXTERIOR TRIM:

Manufacturer

Type

Color

EXTERIOR DOORS:

Manufacturer

Type

Color

WINDOWS:

Manufacturer

Type

Color

*Note to specifier: Suggested paint types and coats for various applications:

*Exterior:

Wood for semi-transparent finish: Semi-transparent stain, 2 coats.
Wood for opaque finish: Alkyd primer; alkyd enamel, 2 coats.
Galvanized metal: Galvanized metal primer; alkyd enamel, 2 coats.
Galvanized metal (high performance): Epoxy primer; catalyzed urethane, 1 coat.
Ferrous metal: Zinc chromate primer; alkyd enamel, 2 coats.
Ferrous metal (high performance): Zinc rich primer, epoxy, 1 coat; catalyzed urethane, 1 coat.
Concrete, stucco and masonry: Acrylic latex, 2 coats.
Concrete masonry units: Block filler; acrylic latex, 2 coats.

INTERIOR PAINTING SCHEDULE

*Note to specifier: List first, second and finish coats by color and gloss:

               Room   Walls   Trim   Ceiling   Floor

COAT 1

Manufacturer

Type

Color

COAT 2

Manufacturer

Type

Color

COAT 1

Manufacturer

Type

Color

*Note to specifier: Suggested paint types and coats for various applications:

*Interior:

Wood for transparent finish: Oil stain; sanding sealer; alkyd varnish, 2 coats.
Wood for opaque finish: Alkyd enamel undercoat; alkyd enamel, 2 coats.
Drywall and plaster: Latex primer; acrylic latex (eggshell), 2 coats.
Drywall and plaster (high performance): Latex primer; polychromatic vinyl copolymer.
Drywall and plaster (heavy duty): Latex primer; water-based epoxy, 2 coats.
Ferrous metal: Alkyd metal primer; alkyd enamel, 2 coats.
Ferrous metal (high performance): Epoxy primer; catalyzed urethane, 2 coats.

2.2 MATERIALS -- LIMITATIONS ON USE

A. Undercoats and thinners:

_ Undercoat must be from the same manufacturer as the finish coat.
_ Thinners must be as recommended by the paint manufacturer and used only as recommended.
_ The undercoat, finish coat and thinner are integrated parts of a total paint finish.

B. Primers:

_ Do not use latex primer on bare wood.
_ Do not use alkyd primer on gypsum board.
_ Apply primer or sealer to knots and pitch pockets on wood that is to be painted.

2.3 MATERIALS DELIVERY AND STORAGE

A. Delivery:

_ All paint materials shall be delivered:

_ New _ In labeled, unopened containers

_ Material quality shall be:

_ Verified as necessary by onsite tests.
_ Verified as necessary by laboratory tests.

B. Storage:

_ Do not use mixed brands or partial substitutions.
_ Have materials delivered in a timely sequence as required to expedite the work flow.

_ Store all paint materials:

_ With ample ventilation.
_ In fire-protected space.
_ Secure from damage.

_ Keep paint storage areas clean and clear of:

_ Spilled material.
_ Empty containers.
_ Rags and scrap.

2.4 APPLICATION EQUIPMENT

A. Use painting tools and equipment only as recommended by the paint manufacturer.

_ Prior to work, verify that proposed equipment is compatible with material to be applied.

2.5 WORKING CONDITIONS

A. Maintain a proper work environment:

_ Dry _ Clean _ Well ventilated
_ Free of airborne construction dust
_ Well lighted
_ In temperature range required by paint manufacturer

_ Keep humidity low enough to prevent moisture condensation on work surfaces.
_ Do not apply paint in snow, rain, or fog; or when relative humidity exceeds 85 percent
_ Never apply paint to damp or wet surfaces.

2.6 MATERIALS PREPARATION

A. Handle and mix paint materials strictly according to manufacturers' instructions.

_ Store paint materials in tightly covered containers when not in use.
_ Maintain paint storage and mix containers clean and free from dirt or paint residue.
_ Stir materials as required to produce a completely uniform mix.
_ Remove and strain away any paint film.
_ Never stir paint film into the mix.

2.7 SURFACE PREPARATION

A. Work preparation:

_ Test areas with paint, and match dried paint to approved color and texture samples.
_ Colors and application:
_ Keep color samples on hand, and use them continuously for comparisons.
_ Use completed color schedule.

B. Prepare and clean working surfaces as per paint manufacturers' instructions.

_ Remove or protect items attached to work surfaces which are not to be painted.
_ After painting in each area, reinstall removed items using workers competent in the related trades.
_ Remove oil and grease with clean cloths.
_ Cleaning must not contaminate adjacent freshly-painted surfaces.
_ Cleaning solvent must meet safety standards of governing building and safety codes.
_ Cleaned surfaces must be thoroughly dry before painting.

_ Thoroughly clean surfaces of construction droppings or stains such as:

_ Rust _ Mortar _ Sealants _ Waterproofing

C. Preparation of wood surfaces:

_ Clean wood of dirt, oil, and any other material that may interfere with painting.
_ Sand exposed wood to smooth uniform surface.
_ Do not paint wood having moisture content of 12% or higher.
_ Measure moisture content of wood with an approved moisture meter.

D. Preparation of metal surfaces:

_ Clean metal of dirt, oil, and any other material that might interfere with painting
_ Use solvent to start cleaning galvanized surfaces.
_ Treat galvanized surface with phosphoric acid etch.
_ Remove etching solution completely before proceeding.

PART 3 -- APPLICATION

3.1 PREPARATION AND COORDINATION:

A. Preparation and coordination:

_ Handle and store materials as per manufacturer's instructions.
_ Maintain a painting log to control schedule and completeness of painting.
_ Remove or fully protect adjacent or related work that might be marred by painting.

_ Coordinate painting with adjacent or related work such as:

_ Hardware   _ Electrical and plumbing fixtures
_ Outlets and switch boxes   _ Trim

B. Coordination with pre-primed or other materials:

_ Touch up and repair any damaged shop-applied prime coats.
_ Touch up bare areas prior to start of finish coats application.
_ Finish coat materials must be compatible with prime coats.
_ Review other Sections of these Specifications to determine prime coatings on various materials.
_ Notify the Architect in writing of anticipated problems with painting over previous prime coats.
_ Do not allow paint gaps or overlaps at edges of hardware, fixtures, or trim.

3.2 PAINT APPLICATION

A. Mix and apply materials as per manufacturer's instructions:

_ Apply coats as per manufacturer's instructions:

_ Thicknesses   _ Curing time between coats
_ Numbers of coats
_ Varied tints for each coat   _ Paint thickness meter testing

_ Apply paint to thoroughly cover undercoat, and do not allow:

_ Show-through _ Lumps _ Runs _ Droplets
_ Lap or brush marks
_ Ripples _ Streaks

_ Do not allow variation in a coat application in:

_ Color _ Texture _ Finish _ Light reflectance

_ Vary the hue of succeeding coats slightly to clearly show coats are applied as required.
_ Do not apply added coats until completed coat is inspected and approved.
_ Only coats of paint inspected and approved by the Architect will be counted as completed.
_ Sand defects smooth between coats.
_ Defects are defined as irregularities visible to the unaided eye at a five foot distance.

B. Keep approved samples on hand for comparison with work.

C. Drying, brushing, and spraying:

_ Allow drying time between coats as instructed by the paint manufacturer.
_ Work and smooth out brush coats onto surface in an even film.
_ Where spraying, apply each coat to provide the hiding equivalent of brush coats.
_ Do not double back with spray equipment to build up film thickness of two coats in one pass.
_ Match applied work with approved samples as to texture, color, and coverage.

3.3 PAINTING AND CLEANING SPECIAL SURFACES:

A. Exposed mechanical items:

_ Paint registers, panels, access doors, ducts, etc. to match adjacent surfaces.
_ Paint back sides of access panels to match exposed sides.
_ Paint visible duct surfaces behind vents, registers, and grilles as directed by the Architect.
_ Exposed vents: Apply two coats of heat-resistant paint.

B. Cleaning and prime coats on metal:

_ Wash metal with approved solvent.
_ Add prime coat followed by two coats of alkyd enamel.

C. Exposed pipe and duct insulation:

_ Apply one coat of latex on insulation which has been sized or primed under another Section.
_ Apply two coats on such surfaces when unprimed.
_ Match color of adjacent surfaces.
_ Remove pipe or duct bands before painting, and replace after painting.

D. Hardware:

_ Paint prime-coated hardware to match adjacent surfaces.
_ Allow no paint to come in contact with hardware that is not to be painted.

E. Damp spaces:

_ In shower or toilet rooms and other damp rooms add approved fungicide to paints.
_ Interior: "Stipple" finish for enamel.

3.4 CLEANING AND EXTRA STOCK

A. Protection and cleaning:

_ Maintain thorough dust and dirt control throughout the painting process.
_ Thoroughly protect all surfaces that won't be painted with:

_ Clean drop cloths   _ Masking tape

_ Heavy duty masking plastic.
_ Drop cloths must be free of gaps or rips.
_ Immediately clean any spilled materials.
_ Do not allow dirt or spilled materials to be tracked in a work area or to other work areas.
_ Allow absolutely no paint smears or splatters to remain on adjacent surfaces.
_ Do not allow paint to dry in cans that are being used or on applicators.
_ Paint remote and out-of-reach places:
_ Inside cabinets   _ Under shelves and drawers
_ Hinge edges of doors
_ Top edges of high trim   _ Undersides of low trim
_ Behind fixtures and equipment

B. Upon completion of painting work, deliver to the Owner:

_ Extra stock of 10% or more of each color, type, and gloss of paint used in the work.
_ Tightly seal and clearly label each container with notes on contents and location used.

3.5 INSPECTION, TOUCH UP AND REPAIRS.

A. Tests:

_ Tests will be made at random to confirm paint coat thicknesses.
_ Unacceptable work includes surface imperfections such as:

_ Cloudiness   _ Spotting   _ Laps   _ Brush marks   _ Runs   _ Sags
_ Ropiness

B. Repair:

_ Remove, refinish, or repaint work not in compliance with specified requirements.
_ Replace or repair all non-conforming work.
_ Do repairs and touch-ups so they are undetectable.

END OF SECTION

DIVISION 9
FINISHES

WALL COVERINGS
09550

PART 1 -- GENERAL

*Note to specifier: Include generic introductory PART 1 text from the beginning of Division 9.

1.1 WORK

A. Provide everything required to complete wall coverings shown on the Drawings and specified herein.

B. Provide surface preparation for wall coverings or special finishes.

1.2 SUBMITTALS

A. Submit for approval: _ Samples _ Product data _ Warranty
_ Maintenance data _ Extra stock.

1.3 QUALITY ASSURANCE

A. Provide products of acceptable manufacturers which have been in satisfactory use in similar service for three years. Use experienced installers. Deliver, handle, and store materials in accordance with manufacturer's instructions.

PART 2 -- PRODUCTS AND MATERIALS

2.1 MATERIALS

*Note to specifier: Refer to Finish Schedule and specify materials and manufacturer, such as vinyl wall covering as per Federal Specification CC-W-408 Type 2 medium duty wall covering, manufactured by:

PART 3 - INSTALLATION

3.1 INSTALLATION

A. Prepare surfaces as per manufacturer's recommendations.

_ Store materials to acclimatize them to the work area.
_ Prime and seal substrates.
_ Substrate moisture content must be in limits specified by manufacturer.

B. Application:

- Apply adhesive and install with seams plumb and overlapped
- Double-cut to ensure tight closure.
- No seams allowed within 6" of any corner.
- Remove air bubbles, blisters, wrinkles and any other defects.
- No horizontal seams are permitted.
- Remove excess adhesive immediately;
- Clean walls and protect surfaces from dirt and damage.

C. Repair and touch-up:

- Replace or repair all non-conforming work.
- Do repairs and touch-ups so they are undetectable.

END OF SECTION

Notes:

DIVISION 10

SPECIALTIES
10000

CONTENTS

10000 SPECIALTIES -- SPECIFIABLE ITEMS   297

10000 SPECIALTIES (generic text)   298

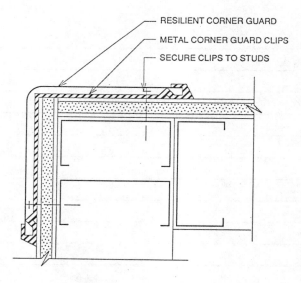

CORNER GUARD

DIVISION 10
SPECIALTIES

SPECIALTIES -- SPECIFIABLE ITEMS
10000

*NOTE TO SPECIFIER:

USE THE GENERIC TEXT FOR PARTS 1, 2, AND 3 FOR ANY
PARTICULAR ITEMS YOU SPECIFY UNDER "DIVISION 10 --
SPECIALTIES."

ADD THE PRODUCT TYPE AND MANUFACTURER, AND REFER
TO THE MANUFACTURER'S INSTRUCTIONS AND SPECIFICATIONS.
IN MOST CASES THIS WILL COVER ALL THE DATA YOU NEED TO
SPECIFY.

Specialties and their CSI numbers include:

_ 10100	Visual Display Boards
	(chalkboards, markerboards, tackboards, etc.)
_ 10150	Toilet Compartments
_ 10185	Shower Cubicles
_ 10200	Louvers and Vents
_ 10240	Grilles and Screens
_ 10250	Service Wall Systems
_ 10260	Wall and Corner Guards
_ 10270	Access Flooring
_ 10290	Pest Control
_ 10300	Fireplaces and Stoves
_ 10340	Manufactured Exterior Specialties
	(steeples, spires, cupolas, etc.)
_ 10350	Flagpoles
_ 10400	Identifying Devices (Directories/Plaques/Signs)
_ 10450	Pedestrian Control Devices
_ 10500	Lockers
_ 10522	Fire Extinguishers, Cabinets, and Accessories
_ 10530	Protective Covers (Walkways/Car Shelters/Awnings)
_ 10550	Postal Specialties
_ 10600	Partitions
_ 10650	Operable Partitions
_ 10670	Storage Shelving
_ 10700	Sun Control Devices (Sun Control/Shutters/Storm Panels)
_ 10750	Telephone Enclosures
_ 10800	Toilet and Bath Accessories and Enclosures
_ 10880	Scales
_ 10900	Wardrobe Specialties

DIVISION 10
SPECIALTIES

SPECIALTIES
10000

PART 1 -- GENERAL

1.1 WORK

A. Provide:

*Note to specifier: Name the Specialties such as any of the items listed on the preceding pages,

Since these are all highly specialized products and systems, you will most likely need to use selected manufacturer's specifications or those from a consultant. You can modify those specifications to include the standard introductory format such as provided below.

B. Provide everything required to complete the work as shown on the Drawings and specified herein.

C. Other related work:

*Note to specifier: This would include reference to related construction that is not provided by the product or system manufacturer such as adjacent excavation, concrete slab supports, etc.

1.2 QUALITY STANDARDS

A. Provide experienced, well-trained workers competent to complete the work as specified.

B. Unless approved by the Architect, provide all related products and accessories from one manufacturer.

C. Use products and accessories:

_ From a manufacturer who specializes in making, installing, and servicing systems of this type.
_ From a manufacturer specified or approved by the Architect.

D. All work shall comply with manufacturer's instructions and governing building and safety codes.

*Note to specifier: Include reference to relevant trade standards, if there are any.

1.3 SUBMITTALS

A. Submit the following within ____ calendar days after receiving the
   Notice to Proceed.

*Note to specifier: Submittals are usually required within a specified number of calendar days after the Contractor is given the Notice to Proceed. 30 calendar days is a common requirement for medium-to-large-size projects. Your choice of time will depend on the size of the project and the Owner's need for an expedited schedule.

_ Submit list of materials to be provided for this work.
_ Submit manufacturer's specifications required to prove compliance with
    these specifications.
_ Submit manufacturer's installation instructions.
_ Submit Shop Drawings as required with complete details and assembly
    instructions.
_ Submit Shop Drawings showing relationship and interface with adjacent
    or related work.
_ Submit samples of proposed exposed finishes and hardware for approval
    by the Architect.

*Note to specifier: Samples may be requested with a different time limit than other submittals. Details of the samples requested -- sizes, finishes, etc. -- are usually specified.

_ At the close of this work, provide two copies of a complete operation
    and maintenance manual.

*Note to Specifier: Use the above note if an operation/maintenance manual is needed and available.

1.4 MATERIALS HANDLING

A. Provide all materials required to complete the work as shown on
   Drawings and specified herein.

_ Deliver, store, and transport materials to avoid damage to the product or
    to any other work.
_ Return any products or materials delivered in a damaged or
    unsatisfactory condition.
_ Materials and products delivered will be certified by the manufacturer to
    be as specified.

B. Store materials in a safe, secure location, protected from dirt, moisture,
   contaminants, and weather.

1.5 PRECONSTRUCTION AND PREPARATION

A. Examine and verify that job conditions are satisfactory for speedy and acceptable work.

_ Maintain and use all up-to-date construction documents on site.
_ Maintain and use up-to-date trade standards and manufacturer's instructions.
_ Confirm there is no conflict between this work and governing building and safety codes.
_ Confirm there are no conflicts between this work and work of other trades.
_ Confirm that work of other trades that must precede this work has been completed.
_ Meet all requirements to secure warranty.

B. Planning and coordination:

_ Notify Architect when work is scheduled to be installed.
_ If required by the Architect, a preconstruction meeting will be held with all concerned parties.
_ Use agreed schedule for installation and for field observation by Architect.

PART 2 -- MATERIALS

2.1 GENERAL

A. Product shall be as manufactured by:

*Note to specifier: Include note "(product) shall be as manufactured by (company name). As appropriate, add product type, trade name, grade, size, and manufacturer's location.

PART 3 -- CONSTRUCTION AND INSTALLATION

3.1 WORK CONDITIONS

A. Correct any conditions not in compliance with Section 1.5.A. noted above.

B. Correct any conditions that might interfere with speedy, well-coordinated execution of the work.

C. All work conditions shall be as per:

_ Manufacturer's instructions.
_ Trade association standards.
_ Governing building and safety codes.

3.2 PREPARATION

A. *Note to specifier: Follow manufacturer, trade association, or consultant's specifications for this subsection.

3.3 INSTALLATION

A. Install products as per Drawings and these Specifications.

*Note to specifier: Follow manufacturer, trade association, or consultant's specifications for this subsection.

B. Upon completion:

_ Secure all required tests, inspections, and approvals of the completed system.
_ Make all required adjustments and corrections at no added cost to the Owner.

C. Provide for maintenance of this work for one year following final acceptance by Owner.

_ Maintenance includes all work required in manufacturer's instructions:

_ Inspection and adjustment.
_ Repair and replacement of parts as required.

3.4 REPAIR AND CLEANUP

A. After installation, inspect all work for improper installation or damage.

_ Operating hardware must perform smoothly.
_ Repair or replace any defective work.
_ Repair work will be undetectable.
_ Redo repairs if work is still defective, as directed by the Architect.

B. Clean the work area and remove all scrap and excess materials from the site.

END OF SECTION

Notes:

_____
_____
_____
_____
_____
_____
_____

DIVISION 11

EQUIPMENT
11000

CONTENTS

11000 EQUIPMENT -- SPECIFIABLE ITEMS	303
11000 EQUIPMENT (generic text)	304

DIVISION 11
EQUIPMENT

EQUIPMENT -- SPECIFIABLE ITEMS
11000

*NOTE TO SPECIFIER:

USE THE GENERIC TEXT FOR PARTS 1, 2, AND 3 FOR ANY PARTICULAR ITEMS YOU SPECIFY UNDER "DIVISION 11 -- EQUIPMENT."

ADD THE PRODUCT TYPE AND MANUFACTURER, AND REFER TO THE MANUFACTURER'S INSTRUCTIONS AND SPECIFICATIONS. IN MOST CASES THIS WILL COVER ALL THE DATA YOU NEED TO SPECIFY.

Items that might be specified in Division 11 include:

- _ 11010 Maintenance Equipment
- _ 11020 Security and Vault Equipment
- _ 11030 Teller and Service Equipment
- _ 11040 Ecclesiastical Equipment
- _ 11050 Library Equipment
- _ 11060 Theater and Stage Equipment
- _ 11090 Checkroom Equipment
- _ 11100 Mercantile Equipment
- _ 11120 Vending Equipment
- _ 11130 Audio-Visual Equipment
- _ 11140 Vehicle Service Equipment
- _ 11150 Parking Control Equipment
- _ 11160 Loading Dock Equipment
- _ 11170 Waste Handling Equipment
- _ 11300 Fluid Waste Treatment and Disposal Equipment
- _ 11400 Food Service Equipment
- _ 11450 Residential Equipment
  - _ Kitchen
  - _ Laundry

- _ 11460 Unit Kitchens and Cabinets
- _ 11470 Darkroom Equipment
- _ 11480 Athletic Equipment
- _ 11500 Industrial and Process Equipment
- _ 11600 Laboratory Equipment
- _ 11680 Office Equipment
- _ 11700 Medical Equipment
- _ 11780 Mortuary Equipment

DIVISION 11
EQUIPMENT

EQUIPMENT
11000

PART 1 -- GENERAL

1.1 WORK

A. Provide:

*Note to specifier: Name the Equipment such as any of the items listed on the preceding pages, Since these are all highly specialized products and systems, you will most likely need to use selected manufacturer's specifications or those from a consultant. You can modify those specifications to include the standard introductory format such as provided below.

B. Provide everything required to complete the work as shown on the Drawings and specified herein.

C. Other related work:

*Note to specifier: This would include reference to related construction that is not provided by the product or system manufacturer such as adjacent excavation, concrete slab supports, etc.

1.2 QUALITY STANDARDS

A. Provide experienced, well-trained workers competent to complete the work as specified.

B. Unless approved by the Architect, provide all related products and accessories from one manufacturer.

C. Use products and accessories:

_ From a manufacturer who specializes in making, installing, and servicing systems of this type.
_ From a manufacturer specified or approved by the Architect.

D. All work shall comply with manufacturer's instructions and governing building and safety codes.

*Note to specifier: Include reference to relevant trade standards, if there are any.

1.3 SUBMITTALS

A. Submit the following within _____ calendar days after receiving the Notice to Proceed.

*Note to specifier: Submittals are usually required within a specified number of calendar days after the Contractor is given the Notice to Proceed. 30 calendar days is a common requirement for medium-to-large-size projects. Your choice of time will depend on the size of the project and the Owner's need for an expedited schedule.

_ Submit list of materials to be provided for this work.
_ Submit manufacturer's specifications required to prove compliance with these specifications.
_ Submit manufacturer's installation instructions.
_ Submit Shop Drawings as required with complete details and assembly instructions.
_ Submit Shop Drawings showing relationship and interface with adjacent or related work.
_ Submit samples of proposed exposed finishes and hardware for approval by the Architect.

*Note to specifier: Samples may be requested with a different time limit than other submittals. Details of the samples requested -- sizes, finishes, etc. -- are usually specified.

_ At the close of this work, provide two copies of a complete operation and maintenance manual.

*Note to Specifier: Use the above note if an operation/maintenance manual is needed and available.

1.4 MATERIALS HANDLING

A. Provide all materials required to complete the work as shown on Drawings and specified herein.

_ Deliver, store, and transport materials to avoid damage to the product or to any other work.
_ Return any products or materials delivered in a damaged or unsatisfactory condition.
_ Materials and products delivered will be certified by the manufacturer to be as specified.

B. Store materials in a safe, secure location, protected from dirt, moisture, contaminants, and weather.

1.5 PRECONSTRUCTION AND PREPARATION

A. Examine and verify that job conditions are satisfactory for speedy and acceptable work.

_ Maintain and use all up-to-date construction documents on site.
_ Maintain and use up-to-date trade standards and manufacturer's instructions.
_ Confirm there is no conflict between this work and governing building and safety codes.
_ Confirm there are no conflicts between this work and work of other trades.
_ Confirm that work of other trades that must precede this work has been completed.
_ Meet all requirements to secure warranty.

B. Planning and coordination:

_ Notify Architect when work is scheduled to be installed.
_ If required by the Architect, a preconstruction meeting will be held with all concerned parties.
_ Use agreed schedule for installation and for field observation by Architect.

PART 2 -- MATERIALS

2.1 GENERAL

A. Product shall be as manufactured by:

*Note to specifier: Include note "(product) shall be as manufactured by (company name). As appropriate, add product type, trade name, grade, size, and manufacturer's location.

PART 3 -- CONSTRUCTION AND INSTALLATION

3.1 WORK CONDITIONS

A. Correct any conditions not in compliance with Section 1.5.A. noted above.

B. Correct any conditions that might interfere with speedy, well coordinated execution of the work.

C. All work conditions shall be as per:

_ Manufacturer's instructions.
_ Trade association standards.
_ Governing building and safety codes.

3.2 PREPARATION

A. *Note to specifier: Follow manufacturer, trade association, or consultant's specifications for this subsection.

3.3 INSTALLATION

A. Install products as per Drawings and these Specifications.

*Note to specifier: Follow manufacturer, trade association, or consultant's specifications for this subsection.

B. Upon completion:

_ Secure all required tests, inspections, and approvals of the completed system.
_ Make all required adjustments and corrections at no added cost to the Owner.

C. Provide for maintenance of this work for one year following final acceptance by Owner.

_ Maintenance includes all work required in manufacturer's instructions:

_ Inspection and adjustment.
_ Repair and replacement of parts as required.

3.4 REPAIR AND CLEANUP

A. After installation, inspect all work for improper installation or damage.

_ Operating hardware must perform smoothly.
_ Repair or replace any defective work.
_ Repair work will be undetectable.
_ Redo repairs if work is still defective, as directed by the Architect.

B. Clean the work area and remove all scrap and excess materials from the site.

END OF SECTION

Notes:

_____

_____

_____

_____

_____

_____

Notes:

DIVISION 12

FURNISHINGS
12000

CONTENTS

12000 FURNISHINGS -- SPECIFIABLE ITEMS   310

12000 FURNISHINGS (generic text)   311

DIVISION 12
FURNISHINGS

FURNISHINGS -- SPECIFIABLE ITEMS
12000

*NOTE TO SPECIFIER:

USE THE GENERIC TEXT FOR PARTS 1, 2, AND 3 FOR ANY PARTICULAR ITEMS YOU SPECIFY UNDER "DIVISION 12 -- FURNISHINGS."

ADD THE PRODUCT TYPE AND MANUFACTURER, AND REFER TO THE MANUFACTURER'S INSTRUCTIONS AND SPECIFICATIONS. IN MOST CASES THIS WILL COVER ALL THE DATA YOU NEED TO SPECIFY.

Items that might be specified in Division 12 include:

12050	Fabrics
12100	Artwork
12300	Manufactured Casework
12500	Window Treatment
12600	Furniture and Accessories
12670	Rugs and Mats
12700	Multiple Seating
12800	Interior Plants and Planters

DIVISION 12
FURNISHINGS

FURNISHINGS
12000

PART 1 -- GENERAL

1.1 WORK

A. Provide:

*Note to specifier: Name the Furnishings such as any of the items listed on the preceding page, Since these are all highly specialized products and systems, you will most likely need to use selected manufacturer's specifications or those from a consultant. You can modify those specifications to include the standard introductory format such as provided below.

B. Provide everything required to complete the work as shown on the Drawings and specified herein.

C. Other related work:

*Note to specifier: This would include reference to related construction that is not provided by the product or system manufacturer such as adjacent excavation, concrete slab supports, etc.

1.2 QUALITY STANDARDS

A. Provide experienced, well-trained workers competent to complete the work as specified.

B. Unless approved by the Architect, provide all related products and accessories from one manufacturer.

C. Use products and accessories:

_ From a manufacturer who specializes in making, installing, and servicing systems of this type.
_ From a manufacturer specified or approved by the Architect.

D. All work shall comply with manufacturer's instructions and governing building and safety codes.

*Note to specifier: Include reference to relevant trade standards, if there are any.

1.3 SUBMITTALS

A. Submit the following within _____ calendar days after receiving the Notice to Proceed.

*Note to specifier: Submittals are usually required within a specified number of calendar days after the Contractor is given the Notice to Proceed. 30 calendar days is a common requirement for medium-to-large-size projects. Your choice of time will depend on the size of the project and the Owner's need for an expedited schedule.

_ Submit list of materials to be provided for this work.
_ Submit manufacturer's specifications required to prove compliance with these specifications.
_ Submit manufacturer's installation instructions.
_ Submit Shop Drawings as required with complete details and assembly instructions.
_ Submit Shop Drawings showing relationship and interface with adjacent or related work.
_ Submit samples of proposed exposed finishes and hardware for approval by the Architect.

*Note to specifier: Samples may be requested with a different time limit than other submittals. Details of the samples requested -- sizes, finishes, etc. -- are usually specified.

_ At the close of this work, provide two copies of a complete operation and maintenance manual.

*Note to Specifier: Use the above note if an operation/maintenance manual is needed and available.

1.4 MATERIALS HANDLING

A. Provide all materials required to complete the work as shown on Drawings and specified herein.

_ Deliver, store, and transport materials to avoid damage to the product or to any other work.
_ Return any products or materials delivered in a damaged or unsatisfactory condition.
_ Materials and products delivered will be certified by the manufacturer to be as specified.

B. Store materials in a safe, secure location, protected from dirt, moisture, contaminants, and weather.

1.5 PRECONSTRUCTION AND PREPARATION

A. Examine and verify that job conditions are satisfactory for speedy and acceptable work.

_ Maintain and use all up-to-date construction documents on site.
_ Maintain and use up-to-date trade standards and manufacturer's instructions.
_ Confirm there is no conflict between this work and governing building and safety codes.
_ Confirm there are no conflicts between this work and work of other trades.
_ Confirm that work of other trades that must precede this work has been completed.
_ Meet all requirements to secure warranty.

B. Planning and coordination:

_ Notify Architect when work is scheduled to be installed.
_ If required by the Architect, a preconstruction meeting will be held with all concerned parties.
_ Use agreed schedule for installation and for field observation by Architect.

PART 2 -- MATERIALS

2.1 GENERAL

A. Product shall be as manufactured by:

*Note to specifier: Include note "(product) shall be as manufactured by (company name). As appropriate, add product type, trade name, grade, size, and manufacturer's location.

PART 3 -- CONSTRUCTION AND INSTALLATION

3.1 WORK CONDITIONS

A. Correct any conditions not in compliance with Section 1.5.A. noted above.

B. Correct any conditions that might interfere with speedy, well-coordinated execution of the work.

C. All work conditions shall be as per:

_ Manufacturer's instructions.
_ Trade association standards.
_ Governing building and safety codes.

3.2 PREPARATION

A.

*Note to specifier: Follow manufacturer, trade association, or consultant's specifications for this subsection.

3.3 INSTALLATION

A. Install products as per Drawings and these Specifications.

*Note to specifier: Follow manufacturer, trade association, or consultant's specifications for this subsection.

B. Upon completion:

_ Secure all required tests, inspections, and approvals of the completed system.
_ Make all required adjustments and corrections at no added cost to the Owner.

C. Provide for maintenance of this work for one year following final acceptance by Owner.

_ Maintenance includes all work required in manufacturer's instructions:

_ Inspection and adjustment.
_ Repair and replacement of parts as required.

3.4 REPAIR AND CLEANUP

A. After installation, inspect all work for improper installation or damage.

_ Operating hardware must perform smoothly.
_ Repair or replace any defective work.
_ Repair work will be undetectable.
_ Redo repairs if work is still defective, as directed by the Architect.

B. Clean the work area and remove all scrap and excess materials from the site.

END OF SECTION

Notes:

DIVISION 13

SPECIAL CONSTRUCTION
13000

CONTENTS

13000 SPECIAL CONSTRUCTION
-- SPECIFIABLE ITEMS 316

13000 SPECIAL CONSTRUCTION (generic text) 298

DIVISION 13
SPECIAL CONSTRUCTION

SPECIAL CONSTRUCTION -- SPECIFIABLE ITEMS
13000

*NOTE TO SPECIFIER:

USE THE GENERIC TEXT FOR PARTS 1, 2, AND 3 FOR ANY PARTICULAR ITEMS YOU SPECIFY UNDER "DIVISION 13 -- SPECIAL CONSTRUCTION."

ADD THE PRODUCT TYPE AND MANUFACTURER, AND REFER TO THE MANUFACTURER'S INSTRUCTIONS AND SPECIFICATIONS. IN MOST CASES THIS WILL COVER ALL THE DATA YOU NEED TO SPECIFY.

Items that might be specified in Division 13 include:

_ 13010   Air-Supported Structures
_ 13020   Integrated Ceilings
_ 13030   Special-Purpose Rooms and Buildings
          _ Shelters & Booths
          _ Prefab Rooms
          _ Saunas
          _ Steam Baths
          _ Vaults

_ 13080   Sound, Vibration, and Seismic Control
_ 13090   Radiation Protection
_ 13120   Pre-Engineered Structures
          _ Metal Building Systems
          _ Greenhouses
          _ Portable and Mobile Buildings

_ 13152   Swimming Pools
_ 13160   Aquariums
_ 13170   Hot Tubs, Whirlpool Tubs
_ 13175   Ice Rinks
_ 13180   Site-Constructed Incinerators
_ 13185   Kennels and Animal Shelters
_ 13200   Liquid and Gas Storage Tanks
_ 13600   Solar Energy Systems
_ 13700   Wind Energy Systems
_ 13800   Building Automation Systems
_ 13900   Fire-Suppression Systems
_ 13950   Special Security Systems

DIVISION 13
SPECIAL CONSTRUCTION

SPECIAL CONSTRUCTION
13000

PART 1 -- GENERAL

1.1 WORK

A. Provide:

*Note to specifier: Name the Special Construction such as any of the items listed on the preceding page, Since these are all highly specialized products and systems, you will most likely need to use selected manufacturer's specifications or those from a consultant. You can modify those specifications to include the standard introductory format such as provided below.

B. Provide everything required to complete the work as shown on the Drawings and specified herein.

C. Other related work:

*Note to specifier: This would include reference to related construction that is not provided by the product or system manufacturer such as adjacent excavation, concrete slab supports, etc.

1.2 QUALITY STANDARDS

A. Provide experienced, well-trained workers competent to complete the work as specified.

B. Unless approved by the Architect, provide all related products and accessories from one manufacturer.

C. Use products and accessories:

_ From a manufacturer who specializes in making, installing, and servicing systems of this type.
_ From a manufacturer specified or approved by the Architect.

D. All work shall comply with manufacturer's instructions and governing building and safety codes.

*Note to specifier: Include reference to relevant trade standards, if applicable.

1.3 SUBMITTALS

A. Submit the following within _____ calendar days after receiving the Notice to Proceed.

*Note to specifier: Submittals are usually required within a specified number of calendar days after the Contractor is given the Notice to Proceed. 30 calendar days is a common requirement for medium-to-large-size projects. Your choice of time will depend on the size of the project and the Owner's need for an expedited schedule.

_ Submit list of materials to be provided for this work.
_ Submit manufacturer's specifications required to prove compliance with these specifications.
_ Submit manufacturer's installation instructions.
_ Submit Shop Drawings as required with complete details and assembly instructions.
_ Submit Shop Drawings showing relationship and interface with adjacent or related work.
_ Submit samples of proposed exposed finishes and hardware for approval by the Architect.

*Note to specifier: Samples may be requested with a different time limit than other submittals. Details of the samples requested -- sizes, finishes, etc. -- are usually specified.

_ At the close of this work, provide two copies of a complete operation and maintenance manual.

*Note to Specifier: Use the above note if an operation/maintenance manual is needed and available.

1.4 MATERIALS HANDLING

A. Provide all materials required to complete the work as shown on Drawings and specified herein.

_ Deliver, store, and transport materials to avoid damage to the product or to any other work.
_ Return any products or materials delivered in a damaged or unsatisfactory condition.
_ Materials and products delivered will be certified by the manufacturer to be as specified.

B. Store materials in a safe, secure location, protected from dirt, moisture, contaminants, and weather.

1.5 PRECONSTRUCTION AND PREPARATION

A. Examine and verify that job conditions are satisfactory for speedy and acceptable work.

_ Maintain and use all up-to-date construction documents on site.
_ Maintain and use up-to-date trade standards and manufacturer's instructions.
_ Confirm there is no conflict between this work and governing building and safety codes.
_ Confirm there are no conflicts between this work and work of other trades.
_ Confirm that work of other trades that must precede this work has been completed.
_ Meet all requirements to secure warranty.

B. Planning and coordination:

_ Notify Architect when work is scheduled to be installed.
_ If required by the Architect, a preconstruction meeting will be held with all concerned parties.
_ Use agreed schedule for installation and for field observation by Architect.

PART 2 -- MATERIALS

2.1 GENERAL

A. Product shall be as manufactured by:

*Note to specifier: Include note "(product) shall be as manufactured by (company name). As appropriate, add product type, trade name, grade, size, and manufacturer's location.

PART 3 -- CONSTRUCTION AND INSTALLATION

3.1 WORK CONDITIONS

A. Correct any conditions not in compliance with Section 1.5.A. noted above.

B. Correct any conditions that might interfere with speedy, well-coordinated execution of the work.

C. All work conditions shall be as per:

_ Manufacturer's instructions.
_ Trade association standards.
_ Governing building and safety codes.

3.2 PREPARATION

A. *Note to specifier: Follow manufacturer, trade association, or consultant's specifications for this subsection.

3.3 INSTALLATION

A. Install products as per Drawings and these Specifications.

*Note to specifier: Follow manufacturer, trade association, or consultant's specifications for this subsection.

B. Upon completion:

_ Secure all required tests, inspections, and approvals of the completed system.
_ Make all required adjustments and corrections at no added cost to the Owner.

C. Provide for maintenance of this work for one year following final acceptance by Owner.

_ Maintenance includes all work required in manufacturer's instructions:

_ Inspection and adjustment.
_ Repair and replacement of parts as required.

3.4 REPAIR AND CLEANUP

A. After installation, inspect all work for improper installation or damage.

_ Operating hardware must perform smoothly.
_ Repair or replace any defective work.
_ Repair work will be undetectable.
_ Redo repairs if work is still defective, as directed by the Architect.

B. Clean the work area and remove all scrap and excess materials from the site.

END OF SECTION

Notes:

_____
_____
_____
_____
_____
_____

DIVISION 14
CONVEYING

CONVEYING
14000

CONTENTS

14000 CONVEYING -- SPECIFIABLE ITEMS	322
14000 CONVEYING (generic text)	323

DIVISION 14
CONVEYING

CONVEYING -- SPECIFIABLE ITEMS
14000

*NOTE TO SPECIFIER:

USE THE GENERIC TEXT FOR PARTS 1, 2, AND 3 FOR ANY PARTICULAR ITEMS YOU SPECIFY UNDER "DIVISION 14 -- CONVEYING."

ADD THE PRODUCT TYPE AND MANUFACTURER, AND REFER TO THE MANUFACTURER'SINSTRUCTIONS AND SPECIFICATIONS. IN MOST CASES THIS WILL COVER ALL THE DATA YOU NEED TO SPECIFY.

Items that might be specified in Division 14 include:

- _ 14100   Dumbwaiters
- _ 14200   Elevators
- _ 14300   Moving Stairs and Walks
- _ 14400   Lifts
- _ 14500   Material Handling Systems
- _ 14600   Hoists and Cranes
- _ 14700   Turntables
- _ 14800   Scaffolding
- _ 14900   Transportation Systems

Standards for common conveying devices: ANSI Safety Code for Elevators, Dumbwaiters, Escalators, and Moving Walks.

DIVISION 14
CONVEYING

CONVEYING
14000

PART 1 -- GENERAL

1.1 WORK

A. Provide:

*Note to specifier: Name the conveying system such as dumbwaiter, electric traction elevator, wheelchair lift, hydraulic elevator, escalator, moving walkway, etc. Since this technology is outside the expertise of most architects, you will most likely need to use selected manufacturer's specifications or those from a consultant. You can modify those specifications to include the standard introductory format such as provided below.

B. Provide everything required to complete the work as shown on the Drawings and specified herein.

C. Other related work:

*Note to specifier: This would include reference to related construction not provided by the conveyor manufacturer or conveyor subcontractor, such as, in the case of an elevator, the elevator pit, hoistway, hoistway structure and machine room.

1.2 QUALITY STANDARDS

A. Provide experienced, well-trained workers competent to complete the work as specified.

B. Unless approved by the Architect, provide all related products and accessories from one manufacturer.

C. Use products and accessories:

_ From a manufacturer who specializes in making, installing, and servicing systems of this type.
_ From a manufacturer specified or approved by the Architect.

D. All work shall comply with manufacturer's instructions and governing building and safety codes.

*Note to specifier: Include relevant trade standards. Elevator and similar products shall comply with ANSI A17, "Safety Code for Elevators, Dumbwaiters, Escalators, and Moving Walks."

1.3 SUBMITTALS

A. Submit the following within _____ calendar days after receiving the Notice to Proceed.

*Note to specifier: Submittals are usually required within a specified number of calendar days after the Contractor is given the Notice to Proceed. 30 calendar days is a common requirement for medium- to large-size projects. Your choice of time will depend on the size of the project and the Owner's need for an expedited schedule.

_ Submit list of materials to be provided for this work.
_ Submit manufacturer's specifications required to prove compliance with these specifications.
_ Submit manufacturer's installation instructions.
_ Submit Shop Drawings as required with complete details and assembly instructions.
_ Submit Shop Drawings showing relationship and interface with adjacent or related work.
_ Submit samples of proposed exposed finishes and hardware for approval by the Architect.

*Note to specifier: Samples may be requested with a different time limit than other submittals. Details of the samples requested -- sizes, finishes, etc. -- are usually specified.

_ At the close of this work, provide three copies of a complete operation and maintenance manual.

1.4 MATERIALS HANDLING

A. Provide all materials required to complete the work as shown on Drawings and specified herein.

_ Deliver, store, and transport materials to avoid damage to the product or to any other work.
_ Reject and return any products or materials delivered in a damaged or unsatisfactory condition.
_ Materials and products delivered will be certified by the manufacturer to be as specified.

B. Store materials indoors, protected from dirt, moisture, contaminants, and weather.

1.5 PRECONSTRUCTION AND PREPARATION

A. Examine and verify that job conditions are satisfactory for speedy and acceptable work.

_ Maintain and use up-to-date construction documents on site.

_ Maintain and use up-to-date trade standards and manufacturer's instructions.
_ Confirm there is no conflict between this work and governing building and safety codes.
_ Confirm there are no conflicts between this work and work of other trades.
_ Confirm that work of other trades that must precede this work has been completed.
_ Meet all requirements to secure warranty.

B. Planning and coordination:

_ Notify Architect when work is scheduled to be installed.
_ If required by the Architect, a preconstruction meeting will be held with all concerned parties.
_ Use agreed schedule for installation and for field observation by Architect.

PART 2 -- MATERIALS

2.1 GENERAL

A. *Note to specifier: Include note "(product) shall be as manufactured by (company name). As appropriate, add product type, trade name, grade, size, and manufacturer's location.

PART 3 -- CONSTRUCTION AND INSTALLATION

3.1 WORK CONDITIONS

A. Correct any conditions not in compliance with Section 1.5.A. noted above.

B. Correct any conditions that might interfere with speedy, well-coordinated execution of the work.

C. All work conditions shall be as per:

_ Manufacturer's instructions.
_ Trade association standards.
_ Governing building and safety codes.

3.2 PREPARATION

A. Shafts, hoistways, and other required construction must be approved:

_ By the conveyor manufacturer.
_ Building department and governing safety regulatory agency.
_ The Architect.

3.3 INSTALLATION

A. Install products as per Drawings and these Specifications.

B. Upon completion:

_ Secure all required tests, inspections, and approvals of the completed system.
_ Make all required adjustments and corrections at no added cost to the Owner.

C. Provide for maintenance of this work for one year following final approval by governing agencies.

_ Maintenance includes all work required in manufacturer's instructions:

_ Inspection, adjustment, lubrication.
_ Repair and replacement of parts as required.
_ Emergency call-back service at all times.

3.4 REPAIR AND CLEANUP

A. After installation, inspect all work for improper installation or damage:

_ Operating hardware must perform smoothly.
_ Repair or replace any defective work.
_ Repair work will be undetectable.
_ Redo repairs if work is still defective, as directed by the Architect.

B. Clean the work area and remove all scrap and excess materials from the site.

END OF SECTION

Notes:

___
___
___
___
___
___
___

Notes:

DIVISION 15

MECHANICAL
15000

CONTENTS

15300	FIRE PROTECTION	328
15400	PLUMBING	333
15500	HEATING, VENTILATION AND AIR CONDITIONING	339

DIVISION 15
MECHANICAL

FIRE PROTECTION
15300

PART 1 -- GENERAL

1.1 WORK

A. Provide and install automatic fire protection sprinkler system as shown on the Drawings and as specified.

B. Fire sprinkler system includes piping, valves, sprinkler heads, water motor and alarm, pressure gauges, hangers and complete support system.

1.2 QUALITY STANDARDS

A. Provide experienced, well-trained workers competent to complete the work as specified.

B. Unless approved by the Architect, provide related products and accessories from one manufacturer.

C. All work shall comply with manufacturer's instructions and governing building and safety codes.

D. All work must be as directed and approved by the governing Fire Rating Bureau, local Fire Marshal, and Standards for Sprinkler System Installations of the National Fire Protection Association.

1.3 SUBMITTALS

A. Submit the following within calendar days after receiving the Notice to Proceed.

*Note to specifier: Submittals are usually required within a specified number of calendar days after the Contractor is given the Notice to Proceed. Your choice of time will depend on the size of the project and the Owner's need for an expedited schedule.

Submit list of materials to be provided for this work.

Submit manufacturer's specifications required to prove compliance with these specifications.

Submit manufacturer's installation instructions.

Submit Shop Drawings and installation drawings as required with complete details and assembly instructions.

Submit Shop Drawings and installation drawings showing relationships and interfaces with adjacent or related work such as air diffusers, lighting fixtures, and structural supports.

Submit samples of proposed fixtures for approval by the Architect.

*Note to specifier: Samples may be requested with a different time limit than other submittals. Details of the samples requested such as sizes, finishes, etc., are usually specified.

At the close of this work, provide Record Drawings of installation, Maintenance Manual, and warranty information.

1.4 MATERIALS HANDLING

A. Provide all materials required to complete the work as shown on Drawings and specified herein. Deliver, store, and transport materials to avoid damage to the product or to any other work. Reject and return any products or materials delivered in a damaged or unsatisfactory condition. Materials and products delivered will be certified by the manufacturer to be as specified.

B. Store materials indoors, protected from dirt, moisture, contaminants, and weather.

1.5 PRECONSTRUCTION AND PREPARATION

A. Examine and verify that job conditions are satisfactory for speedy and acceptable work. Maintain and use up-to-date trade standards and manufacturer's instructions.

B. Confirm there is no conflict between this work and governing building and safety codes. Confirm there are no conflicts between this work and work of other trades. Confirm that work of other trades that must precede this work has been completed. Meet all requirements to secure warranty.

C. Notify Architect when work is scheduled to be installed. Use agreed schedule for installation and for field observation by Architect.

PART 2 -- DESIGN AND MATERIALS

2.1 SPRINKLER SYSTEM DESIGN

A. Provide complete system including supply piping, valves, connection to utility main, related construction and the overhead sprinkler system.

B. Sprinkler heads shall exceed building code and fire code minimum requirements in number and position to assure there will be no blind areas or floor areas or walls not reached by water in the event of fire activation of the sprinkler system.

2.2 MATERIALS

*Note to specifier: Sprinkler system design and specifications are usually provided by a qualified consultant or vendor. Those specifications can be included by reference or incorporated in these. If a consultant provides design and specifications, those specifications may be included here or incorporated by reference.

PART 3 -- CONSTRUCTION AND INSTALLATION

3.1 WORK CONDITIONS

A. Correct any conditions not in compliance with Section 1.5.A. noted above.

B. Correct any conditions that might interfere with speedy, well-coordinated execution of the work.

C. All work conditions shall be as per manufacturer's instructions, trade association standards, and governing building and safety codes.

3.2 PREPARATION

A. Support construction for sprinkler pipes and equipment must be as required by the building department.

3.3 INSTALLATION

A. Install products as per Drawings and these Specifications.

B. Conceal pipes in all areas with finish ceilings. In unfinished areas, secure pipes at minimum distance practical below the underside of the above floor or roof.

C. Upon completion, secure all required tests, inspections, and approvals of the completed system. Make all required adjustments and corrections at no added cost to the Owner.

D. Provide for maintenance of this work for one year following final approval by governing agencies. Maintenance includes all work required in manufacturer's instructions such as inspection, adjustment, repair and replacement of parts as required, and emergency call-back service.

3.4 REPAIR AND CLEANUP

A. After installation, inspect all work for improper installation or damage.

B. Operating hardware must perform smoothly. Repair or replace any defective work. Repair work will be undetectable. Redo repairs if work is still defective, as directed by the Architect or governing safety regulatory agency.

C. Clean the work area and remove all scrap and excess materials from the site.

END OF SECTION

Notes:

DIVISION 15
MECHANICAL

PLUMBING
15400

PART 1 -- GENERAL

1.1 WORK

A. Provide and install plumbing as shown on the Drawings and as specified herein.

B. Plumbing includes:

Hot and cold water distribution systems.

Waste drains and vents.

Gas piping.

Plumbing fixtures.

C. Plumbing fixtures shall be as shown on the Drawings and the Plumbing Fixture Schedule. Work includes trim and related construction as required.

1.2 QUALITY STANDARDS

A. Provide experienced, well-trained workers competent to complete the work as specified.

B. Unless approved by the Architect, provide related products and accessories from one manufacturer.

C. All work shall comply with manufacturer's instructions and governing building and safety codes.

1.3 SUBMITTALS

A. Submit the following within calendar days after receiving the Notice to Proceed.

*Note to specifier: Submittals are usually required within a specified number of calendar days after the Contractor is given the Notice to Proceed. 30 calendar days is a common requirement for medium- to large-size projects. Your choice of time will depend on the size of the project and the Owner's need for an expedited schedule.

Submit list of materials to be provided for this work.

Submit manufacturer's specifications required to prove compliance with these specifications.

Submit manufacturer's installation instructions.

Submit Shop Drawings as required with complete details and assembly instructions.

Submit Shop Drawings showing relationship and interface with adjacent or related work.

Submit samples of proposed exposed finishes and fixtures for approval by the Architect.

*Note to specifier: Samples may be requested with a different time limit than other submittals. Details of the samples requested such as sizes, finishes, etc., are usually specified.

At the close of this work, provide copies of manufacturer's installation, maintenance, and warranty information.

1.4 MATERIALS HANDLING

A. Provide all materials required to complete the work as shown on Drawings and specified herein. Deliver, store, and transport materials to avoid damage to the product or to any other work. Reject and return any products or materials delivered in a damaged or unsatisfactory condition. Materials and products delivered will be certified by the manufacturer to be as specified.

B. Store materials indoors, protected from dirt, moisture, contaminants, and weather.

1.5 PRECONSTRUCTION AND PREPARATION

A. Examine and verify that job conditions are satisfactory for speedy and acceptable work. Maintain and use up-to-date trade standards and manufacturer's instructions.

B. Confirm there is no conflict between this work and governing building and safety codes. Confirm there are no conflicts between this work and work of other trades. Confirm that work of other trades that must precede this work has been completed. Meet all requirements to secure warranty.

C. Notify Architect when work is scheduled to be installed. Use agreed schedule for installation and for field observation by Architect.

PART 2 -- MATERIALS

2.1 SANITARY SEWER PIPES

A. Sanitary sewer piping within the building line shall be:

*Note to specifier: Specify selected product such as these listed below. Verify acceptance by local building code. What follows is a typical set of materials for a residence or small commercial project:

PVC pipe as per ASTM D2729, solvent weld joints.

ABS pipe as per ASTM D2680 or D2751, solvent weld joints.

Cast iron pipe, service weight, neoprene gaskets.

Cast iron pipe, hubless, neoprene gaskets, stainless steel clamps.

B. Sanitary sewer piping below grade beyond the building line shall be:

*Note to specifier: Specify selected product such as these listed below. Verify acceptance by local building code. What follows is a typical set of materials for a residence or small commercial project:

Vitrified clay pipe, standard, bell and spigot, neoprene gaskets.

PVC pipe as per ASTM D3033 or D3034, SDR 35. Elastomeric Gaskets as required.

Cast iron pipe, service weight with neoprene gaskets.

2.2 WATER SUPPLY

A. Water supply pipe below grade beyond the building line shall be:

*Note to specifier: Specify selected product such as these listed below. Verify acceptance by local building code. What follows is a typical set of materials for a residence or small commercial project:

Copper tubing, Type K annealed, wrought copper fittings, compression joints.

Cast iron pipe, ductile iron fittings, rubber gaskets, mechanical joints, 3/4 inch diameter rods.

PVC pipe, Schedule 40, minimum 150 psi pressure rating, solvent weld joints.

B. Water supply pipe above grade and within the building line shall be:

*Note to specifier: Specify selected product such as these listed below. Verify acceptance by local building code.

*Note to specifier: What follows is a typical set of materials for a residence or small commercial project:

Copper tubing, Type M, hard drawn, cast brass or wrought copper fittings, soldered joints.

CPVC pipe as per ASTM D2846, CPVC fittings, solvent weld joints.

PB pipe as per ASTM D3309, ASTM F845 PB, or copper fittings, copper compression rings.

2.3 GAS SUPPLY

*Note to specifier: Specify selected product such as these listed below. Verify acceptance by local building code. What follows is a typical set of materials for a residence or small commercial project:

A. Natural gas pipe below grade and beyond the building line shall be steel, Schedule 40, black with polyethylene jacket, welded joints.

B. Natural gas pipe above grade and within the building line shall be steel, Schedule 40 black, malleable iron or forged steel fittings, screwed or welded.

C. Provide clearly marked, easily accessible, and tested shut off valves as required by the building code.

2.4 FUEL SUPPLY

*Note to specifier: Specify selected product such as these listed below. Verify acceptance by local building code. What follows is a typical set of materials for a residence or small commercial project:

A. Fuel oil pipe below grade shall be steel, Schedule 40 black, polyethylene jacket and welded joints.

B. Fuel oil piping above ground shall be steel, Schedule 40 black, malleable iron or forged steel fittings, screwed or welded.

C. Provide clearly marked, easily accessible, and tested shut off valves as required by the building code.

2.5 WATER HEATER

A. Provide and install an automatic water heater as shown on the Drawings.

*Note to specifier: Refer to water heater size, manufacturer, model number, flue passage, baffle and draft hood, insulation, finish, thermostat, pressure regulator, and safety devices. Specify gas or electric powered. Refer to tie downs in earthquake areas or as required by the building code.

2.6 PLUMBING FIXTURES

A. Provide and install plumbing fixtures, trim, and related construction as per the Drawings and Plumbing Fixture Schedule in the Drawings.

*Note to specifier: Refer to a Plumbing Fixture Schedule in the Drawings, or create a Schedule in the Specifications with the following information:

_ Fixture Manufacturer
_ Model
_ Finish/Color
_ Trim/Related Fixtures Size
_ Mounting Height

PART 3 -- CONSTRUCTION AND INSTALLATION

3.1 WORK CONDITIONS

A. Correct any conditions not in compliance with Section 1.5.A. noted previously.

B. Correct any conditions that might interfere with speedy, well-coordinated execution of the work.

C. All work conditions shall be as per manufacturer's instructions, trade association standards, and governing building and safety codes.

3.2 PREPARATION

A. Vents and related support construction for plumbing and mechanical equipment must be as required by the building department.

3.3 INSTALLATION

A. Install products as per Drawings and these Specifications.

B. Upon completion, secure all required pressure tests, inspections, and approvals of the completed system. Make all required adjustments and corrections at no added cost to the Owner. Sterilize the water system and provide copies of a Certificate of Performance.

C. Provide for maintenance of this work for one year following final approval by governing agencies. Maintenance includes all work required in manufacturer's instructions such as inspection, adjustment, repair and replacement of parts as required.

3.4 REPAIR AND CLEANUP

A. After installation, inspect all work for improper installation or damage.

B. Operating fixtures must perform smoothly. Repair or replace any defective work. Repair work will be undetectable. Redo repairs if work is still defective, as directed by the Architect or governing safety regulatory agency.

C. Clean the work area and remove all scrap and excess materials from the site.

END OF SECTION

Notes:

DIVISION 15
MECHANICAL

HEATING, VENTILATION AND AIR CONDITIONING
15500

PART 1 -- GENERAL

1.1 WORK

*Note to specifier: If system is a forced warm air system, note that the work includes furnace, ductwork, and air outlets.

If the system is hydronic, note that the work includes a hot water boiler, hot water pumps, hot water distribution and radiators.

If the system consists of separate self-contained units, describe the units and related support and plumbing/electrical construction.

If the system is solar or includes solar or other components, list each major component as part of the work.

A. Provide and install heating, ventilation, and air conditioning systems as shown on the Drawings and as specified.

B. Mechanical equipment and fittings shall be as shown on the Drawings. Work includes trim and related construction as required.

1.2 QUALITY STANDARDS

A. Provide experienced, well-trained workers competent to complete the work as specified.

B. Unless approved by the Architect, provide related products and accessories from one manufacturer.

C. All work shall comply with manufacturer's instructions and governing building and safety codes.

1.3 SUBMITTALS

A. Submit the following within calendar days after receiving the Notice to Proceed.

*Note to specifier: Submittals are usually required within a specified number of calendar days after the Contractor is given the Notice to Proceed. Your choice of time will depend on the size of the project and the Owner's need for an expedited schedule.

Submit list of materials to be provided for this work.

Submit manufacturer's specifications required to prove compliance with these specifications.

Submit manufacturer's installation instructions.

Submit Shop Drawings as required with complete details and assembly instructions.

Submit Shop Drawings showing relationship and interface with adjacent or related work.

Submit samples of proposed exposed finishes and fixtures for approval by the Architect.

*Note to specifier: Samples may be requested with a different time limit than other submittals. Details of the samples requested such as sizes, finishes, etc., are usually specified.

At the close of this work, provide copies of manufacturer's installation, maintenance, and warranty information.

1.4 MATERIALS HANDLING

A. Provide all materials required to complete the work as shown on Drawings and specified herein. Deliver, store, and transport materials to avoid damage to the product or to any other work. Reject and return any products or materials delivered in a damaged or unsatisfactory condition. Materials and products delivered will be certified by the manufacturer to be as specified.

B. Store materials indoors, protected from dirt, moisture, contaminants, and weather.

1.5 PRECONSTRUCTION AND PREPARATION

A. Examine and verify that job conditions are satisfactory for speedy and acceptable work. Maintain and use up-to-date trade standards and manufacturer's instructions.

B. Confirm there is no conflict between this work and governing building and safety codes. Confirm there are no conflicts between this work and work of other trades. Confirm that work of other trades that must precede this work has been completed. Meet all requirements to secure warranty.

C. Notify Architect when work is scheduled to be installed. Use agreed schedule for installation and for field observation by Architect.

PART 2 -- MATERIALS

2.1 HEATING SYSTEM

A. Furnace shall be:

*Note to specifier: Include note "(product) shall be as manufactured by (company name)." As appropriate, add product type, model number, size, fuel, and materials of construction. Note type such as package unit, self-contained, prewired, etc.

B. Ductwork shall be:

*Note to specifier: Sheet metal ducts, galvanized sheet metal ductwork as per ASHRAE and SMACNA standards. Branch ducts must have manually operated dampers two gauges heavier than the duct sheet metal. Flexible ductwork: Zinc-coated steel helix with seamless vapor barrier and 1" fiberglass insulation. Flexible duct installed strictly as directed by the manufacturer, without sags or kinks.

2.2 COOLING SYSTEM, AIR CONDITIONING

A. Refrigeration or cooling unit shall be:

*Note to specifier: Include note "(product) shall be as manufactured by (company name)." As appropriate, add cooling system type, model number, size, fuel, and materials of construction.

B. Ductwork shall be:

*Note to specifier: Use note as per 2.1.A above.

2.3 FORCED AIR SYSTEM OUTLETS

A. Diffusers and registers shall be located as shown on the Drawings and as approved by the Architect.

B. Diffusers shall be:

*Note to specifier: Include manufacturer, model, type, and operation.

C. Registers shall be:

*Note to specifier: Include manufacturer, model, type, and operation.

D. Exterior exhaust vents shall be:

*Note to specifier: Include manufacturer, model, type, and operation.

E. Exterior fresh air intake shall be:

*Note to specifier: Include manufacturer, model, type, and operation.

PART 3 -- CONSTRUCTION AND INSTALLATION

3.1 WORK CONDITIONS

A. Correct any conditions not in compliance with Section 1.5.A. noted above.

B. Correct any conditions that might interfere with speedy, well-coordinated execution of the work.

C. All work conditions shall be as per manufacturer's instructions, trade association standards, and governing building and safety codes.

3.2 PREPARATION

A. Vents and related support construction for mechanical equipment must be as required by the building department.

3.3 INSTALLATION

A. Install products as per Drawings and these Specifications.

B. Upon completion, secure all required tests, inspections, and approvals of the completed system.
Make all required adjustments and corrections at no added cost to the Owner.

C. Provide for maintenance of this work for one year following final approval by governing agencies. Maintenance includes all work required in manufacturer's instructions such as inspection, adjustment, lubrication, repair and replacement of parts as required, and emergency call-back service.

3.4 REPAIR AND CLEANUP

A. After installation, inspect all work for improper installation or damage.

B. Operating hardware must perform smoothly. Repair or replace any defective work. Repair work will be undetectable. Redo repairs if work is still defective, as directed by the Architect or governing safety regulatory agency.

C. Clean the work area and remove all scrap and excess materials from the site.

END OF SECTION

Notes:

Notes:

DIVISION 16
ELECTRICAL

ELECTRICAL
16000

CONTENTS

16400	ELECTRICAL POWER DISTRIBUTION	345
16500	LIGHTING	350
16700	COMMUNICATIONS	352

DIVISION 16
ELECTRICAL

ELECTRICAL POWER DISTRIBUTION
16400

PART 1 -- GENERAL

1.1 WORK

A. Provide and install complete electrical service, power and lighting as shown on the Drawings and specified herein.

1.2 QUALITY STANDARDS

A. Provide experienced, well-trained workers competent to complete the work as specified.

B. Unless approved by the Architect, provide all related products and accessories from one manufacturer.

C. Use products and accessories from manufacturers who specialize in making, installing, and servicing systems of this type. From a manufacturer specified or approved by the Architect.

D. All work shall comply with manufacturer's instructions and governing building and safety codes.

1.3 SUBMITTALS

A. Submit the following within calendar days after receiving the Notice to Proceed.

*Note to specifier: Submittals are usually required within a specified number of calendar days after the Contractor is given the Notice to Proceed. Your choice of time will depend on the size of the project and the Owner's need for an expedited schedule.

Submit list of materials to be provided for this work.

Submit manufacturer's specifications required to prove compliance with these specifications.

Submit manufacturer's installation instructions.

Submit Shop Drawings as required with complete details and assembly instructions.

Submit Shop Drawings showing relationship and interface with adjacent or related work.

Submit samples of proposed exposed finishes and hardware for approval by the Architect.

*Note to specifier: Samples may be requested with a different time limit than other submittals. Details of the samples requested -- sizes, finishes, etc. -- are usually specified.

At the close of this work, provide three copies of operations and warranty information.

1.4 MATERIALS HANDLING

A. Provide all materials required to complete the work as shown on Drawings and specified herein. Deliver, store, and transport materials to avoid damage to the product or to any other work. Reject and return any products or materials delivered in a damaged or unsatisfactory condition. Materials and products delivered will be certified by the manufacturer to be as specified.

B. Store materials indoors, protected from dirt, moisture, contaminants, and weather.

1.5 PRECONSTRUCTION AND PREPARATION

A. Examine and verify that job conditions are satisfactory for speedy and acceptable work. Maintain and use up-to-date construction documents on site. Maintain and use up-to-date trade standards and manufacturer's instructions.

B. Confirm there is no conflict between this work and governing building and safety codes. Confirm there are no conflicts between this work and work of other trades. Confirm that work of other trades that must precede this work has been completed. Meet all requirements to secure warranty.

C. Notify Architect when work is scheduled to be installed. Use agreed schedule for installation and for field observation by Architect.

PART 2 -- MATERIALS

2.1 GENERAL

A. All materials must be new and of the type and quality specified. Materials must be delivered in labeled, unopened containers. All electrical products must bear the Underwriters Laboratory label.

2.2 TEMPORARY POWER

A. Provide temporary power, power pole, connection to utility, and temporary meter as required for construction. Provide and install permanent electrical meter as building nears completion.

B. After the permanent meter is connected, the Owner will pay metered costs for electrical power.

2.3 ELECTRICAL SERVICE

A. Provide complete electrical service as shown on the drawings and specified herein.

*Note to specifier: What follows is a typical set of materials for a residence or small commercial project:

B. Service entrance cable, copper conductor, 600 volt insulation, type SE. Main distribution panels: NEMA PB 1; circuit breaker type. Provide surface cabinet with screw cover and hinged door. Copper bus and ground bus, 110/220 volts.

C. Underground feeder and branch circuit cable, copper conductor, 600 volt insulation, type UF.

D. Wiring, nonmetallic sheathed cable, size 14 through 4 AWG, copper conductor, 600 volt insulation, type NM.

E. Conduit, junction boxes, and electrical wire connectors shall be as required by the local building code.

F. Circuits will be as diagrammed on Electrical Drawings.

2.4 SWITCHES, RECEPTACLES AND WALL PLATES

A. Provide complete switches, receptacles, wall plates and related materials as shown on the drawings and specified herein.

*Note to specifier: What follows is a typical set of materials for a residence or small commercial project:

B. Wall switches, quiet operating switch rated 20 amperes and 110-220 volts AC. Color and switch type as selected by the Architect. Wall Dimmers, linear slide type, color selected by Architect. Rated for 600 Watts minimum, size as per circuit.

C. Receptacles, Type 5-20 R, plastic face, color as selected by the Architect. Specific purpose receptacles as shown on the Drawings.

D. Exterior weatherproof cover plates shall be gasketed cast metal with hinged gasketed covers.

PART 3 -- CONSTRUCTION AND INSTALLATION

3.1 WORK CONDITIONS

A. Correct any conditions not in compliance with Section 1.5.A. noted above.

B. Correct any conditions that might interfere with speedy, well-coordinated execution of the work.

C. All work conditions shall be as per manufacturer's instructions, trade association standards, and governing building and safety codes.

3.2 PREPARATION

A. Straps and other support construction for electrical equipment must be as required by the building department.

3.3 INSTALLATION

A. Install products as per Drawings and these Specifications.

B. Upon completion, secure all required pressure tests, inspections, and approvals of the completed system. Make all required adjustments and corrections at no added cost to the Owner.

C. Provide for maintenance of this work for one year following final approval by governing agencies. Maintenance includes all work required in manufacturer's instructions such as inspection, adjustment, repair and replacement of parts as required.

3.4 REPAIR AND CLEANUP

A. After installation, inspect all work for improper installation or damage.

B. Operating fixtures must perform smoothly. Repair or replace any defective work. Repair work will be undetectable. Redo repairs if work is still defective, as directed by the Architect or governing regulatory agency.

C. Clean the work area and remove all scrap and excess materials from the site.

END OF SECTION

Notes:

DIVISION 16
ELECTRICAL

LIGHTING
16500

PART 1 -- GENERAL

1.1 WORK

A. Provide and install all interior and exterior lighting fixtures and lamps as shown on the Drawings and specified herein.

*Note to specifier: General items of scope of quality standards, submittals, etc. shall be as per PART 1 -- GENERAL for ELECTRICAL POWER DISTRIBUTION at the beginning of this Division.

PART 2 -- MATERIALS

2.1 GENERAL

A. All materials must be new and of the type and quality specified. Materials must be delivered in labeled, unopened containers. All electrical products must bear the Underwriters Laboratory label.

B. Products and requirements for lamps and lampholders are shown in the Lighting Fixture Schedule.

C. Provide and install all required accessories for mounting and operation of each fixture.

*Note to specifier: If the Lighting Fixture Schedule is included in these specifications, it should list all fixtures, manufacturers, model names or numbers, types, ratings, materials, and finishes.

PART 3 -- CONSTRUCTION AND INSTALLATION

*Note to specifier: This section shall be as per PART 3 CONSTRUCTION AND INSTALLATION for ELECTRICAL POWER DISTRIBUTION in the first section of this Division.

END OF SECTION

Notes:

Notes:

DIVISION 16
ELECTRICAL

COMMUNICATIONS
16700

PART 1 -- GENERAL

1.1 WORK

A. Provide and install all interior and exterior lighting fixtures and lamps as shown on the Drawings and specified herein.

*Note to specifier: General items of scope of quality standards, submittals, etc. shall be as per PART 1 -- GENERAL for ELECTRICAL POWER DISTRIBUTION at the beginning of this Division.

PART 2 -- MATERIALS

2.1 GENERAL

A. All materials must be new and of the type and quality specified. Materials must be delivered in labeled, unopened containers. All electrical products must bear the Underwriters Laboratory label.

2.2 COMMUNICATIONS

A. Provide complete telephone and video cable service as shown on the drawings and specified herein.

B. Telephone service includes equipment not provided by the phone company and may include service entrance equipment, outlets, terminal boards and other items shown on the Drawings or required for a complete operating telephone service.

C. Coordinate installation of items provided by the telephone company.

D. Cable television includes equipment not provided by the cable company and may include service entrance equipment, outlets, and other items shown on the Drawings or required for a complete operating cable service.

E. Coordinate installation of materials provided and installed by the cable company.

PART 3 -- CONSTRUCTION AND INSTALLATION

*Note to specifier: This section shall be as per PART 3 CONSTRUCTION AND INSTALLATION for ELECTRICAL POWER DISTRIBUTION in the first section of this Division.

END OF SECTION

Notes:

SPECIFICATIONS DATA SOURCES --
GUIDE AND MASTER SPECIFICATIONS

A-1

__ "MASTERFORMAT" sections "Tech Aids," and much more.

    CSI -- Construction Specifications Institute
    601 Madison St.
    Alexandria, VA 22314
    703-684-0300
    Fax: 703-684-0465
    www.CSINET.org

__ AIA MASTERSPEC

    AIA Service Corporation
    1735 New York Ave., N.W.
    Washington, DC 20006
    1-800-424-5080
    www.AIA.org

__ Sweet's Building Selection Data and Sweet's catalogs

    McGraw-Hill Information Systems Co.
    1221 Avenue of the Americas
    New York, NY 10020
    212-512-3268

__ QUICK SPECS. Simplified outline master specifications suited to smaller offices and projects (buildings up to $4 million).

    Guidelines
    Box 456
    Orinda, CA 94563
    1-800-634-7779
    Fax: 925-299-0181
    E-mail: FSGL@AOL.com

__ LIBRARY OF SPECIFICATIONS SECTIONS. Over 250 master specification sections covering all 16 CSI divisions. This is an outstanding source of text in four volumes by specifications expert Hans W. "Bill" Meier, FCSI. Prentice-Hall Business & Professional Books, Englewood Cliffs, NJ 07632.

__ NAVFAC GuideSpecifications
    Naval Facilities Engineering Command
    200 Stovall St.
    Alexandria, VA 22332
    703-325-0310
    Guide Specifications Division (Code 15G)
    805-982-5661
    www.navy.mil:80/homepages/navfac/index.html

SPECIFICATIONS DATA SOURCES --
GUIDE AND MASTER SPECIFICATIONS continued

_ MILITARY STANDARDS SPECIFICATIONS
   Defense Automated Printing Service
   Dept. of the Navy
   700 Robbins Ave.
   Philadelphia, PA 19111
   215-697-4742
   Fax: 215-697-2978
   www.DODSSP.daps.mil/
   Department of Defense Index of Specifications and Standards
   www.dtic.mil/stinet/publicstinet/htgi/dodiss/

_ Technical, facilities management, and design standards

   General Services Admnistration
   Public Buildings Service
   1800 F St. N.W., Room 6344
   Washington, DC 20405
   202-501-1100
   www.GSA.gov
   www.GSA.gov/pbs/pbs.htm

_ Technical, facilties management, and design standards

   Veterans Health Administration
   Office of Facilities Management
   Department of Veterans Affairs
   810 Vermont Ave. N.W.
   Washington, DC 20420
   www.va.gov/facmgt/homefm.htm

REFERENCE STANDARDS

_ Publications of the American National Standards Institute

   ANSI -- American National Standards Institute
   11 West 42nd St., 13th floor
   New York, New York 10036
   212-642-4900
   Fax: 212-398-0023
   www.anis.org

_ ASTM Publications Catalog. ASTM.
_ Compilation of ASTM Standards in Building Codes, ASTM.

   ASTM -- American Society for Testing and Materials
   100 Barr Harbor Drive
   W. Conshohocken, PA 19428
   610-832-9500
   Fax: 610-832-9555
   E-mail: service@local.astm.org
   www.ASTM.org

SPECIFICATIONS DATA SOURCES -- TECHNICAL DATA

03000 CONCRETE & PORTLAND CEMENT

__ American Concrete Institute Publications Catalog. ACI.
__ An Architect's Guide to ACI Publications. ACI.

    ACI -- American Concrete Institute
    P. O. Box 9094
    Farmington Hills, MI 48333
    248- 848-3700
    Fax: 248-848-3701

    APA -- Architectural Precast Association
    P.O. Box 08669
    Fort Myers, FL 33908-0669
    941-454-6989
    Fax: 941-454-6787
    E-mail: concrete@water.net

__ Reinforced Concrete, Manual of Standard Practice (annual).

    CRSI -- Concrete Reinforcing Steel Institute
    933 N. Plum Grove Rd.
    Schaumburg, IL 60173
    847-517-1200
    Fax: 847-517-1206
    E-mail: info@CRSI.org
    www.CRSI.org

    NPCA -- National Precast Concrete Association
    10333 North Meridian Street, Suite 272
    Indianapolis, IN 46290-1081
    317-571-9500
    Fax: 317-571-0041

__ Publications of the Portland Cement Association

    PCA -- Portland Cement Association
    5420 Old Orchard Rd.
    Skokie, IL 60077
    847-966-6200
    1-800-868-6733
    Fax: 847-966-8389

    TCA -- Tilt-Up Concrete Association
    121-1/2 1st Street West
    Mount Vernon, IA 52314
    319-895-6911
    Fax: 319-895-8830
    E-mail: esauter@tilt-up.org
    www.TILT-UP.org

SPECIFICATIONS DATA SOURCES -- TECHNICAL DATA continued

04000 MASONRY

__ Brick, block, and stone masonry detailing and specifications information.

BIA -- Brick Institute of America
11490 Commerce Park Drive
Reston, VA 20191
703-620-0010
Fax: 703-620-3928
E-mail: brick@pop.erols.com
www.BRICKINST.org

IMI -- International Masonry Institute
823 15th St. N.W., Suite 1001
Washington, DC 20005
202-783-3908
Fax: 202-783-0438
1-800-464-0988

MIA -- Masonry Institute of America
2550 Beverly Blvd.
Los Angeles, CA 90059
213-388-0472
Fax: 213-389-7514
E-mail: askus@masonryinstitute.org
www.MASONRYINSTITUTE.org

MIA -- Marble Institute of America
30 Eden Alley, Suite 201
Columbus, OH 43215
614-228-6194
Fax: 614-228-7434

NCMA -- National Concrete Masonry Association
2302 Horse Pen Road
Herndon, VA 20171-3499
703-713-1900
Fax: 703-713-1910

SPECIFICATIONS DATA SOURCES --
TECHNICAL DATA continued

05000 METALS & STEEL

__ Manual of Steel Construction. American Institute of Steel Construction.

__ Specifications for the Design, Fabrication and Erection of Structural Steel for Buildings.

    AISC -- American Institute of Steel Construction
    1 East Wacker Drive, Suite 3100
    Chicago, IL 60601
    312-670-2400
    Fax: 312-670-5403
    E-mail: Johnson@aisc.com
    www.AISCWEB.com

    AISI -- American Iron and Steel Institute
    1101 17th Street N.W., Suite 1300
    Washington, D.C. 20036
    202-452-7100
    Fax: 202-463-6573
    www.STEEL.org

__ Standard specifications, technical standards for galvanizing protection of steel.

    AGA -- American Galvanizers Association
    12200 E. Iliff, Suite 204
    Aurora, CO 80014
    303-750-2900
    1-800-468-7732
    Fax: 303-750-2909
    E-mail: aga@netway.net
    www.USALINK.net/aga

__ Technical standards for metal fabrications.

    NAAMM --
    National Association of Architectural Metal Manufacturers
    8 South Michigan Avenue, Suite 1000
    Chicago, IL 60603
    312-332-0405
    Fax: 312-332-0706
    E-mail: naamm@gss.net
    www.gss.net/naamm

    Sheet Metal & Air Conditioning Contractors National Assoc.
    4201 Lafayette Center Drive
    Chantilly, VA 22021
    703-803-2980
    Fax: 703-803-3732
    www.SMACNA.org

SPECIFICATIONS DATA SOURCES -- TECHNICAL DATA continued

A-6

05000 METALS & STEEL continued

__ Standard Specifications, Load Tables, and Weight Tables for Steel Joists and Joist Girders.

SJI -- Steel Joist Institute, 3127 10th Avenue
North Extension,
Myrtle Beach, SC 29577
803-626-1995
Fax: 803-629-5565

06000 WOOD

__ Wood design standards information.

AFPA/AWC -- American Forest & Paper Association/
American Wood Council
1111 19th St. N.W., Suite 800
Washington, DC 20036
202-463-2700
Fax: 202-463-2785
www.AWC.org

__ Design Standard Specifications for Structural Glued Laminated Timber of Softwood Species, 1980

AITC -- The American Institute of Timber Construction
7012 South Revere Parkway, Suite 140
Englewood, CO 80112
303-792-9559
Fax: 303-792-0669

__ Hardwood design standards information.

AHA -- American Hardboard Association
1210 W. Northwest Highway
Palatine, IL 60067
847-934-8800
Fax: 847-934-8803

__ Design standards for plywood, waferboard, strandboard, etc.

APA -- The Engineered Wood Association
P.O. Box 11700
Tacoma, WA 98411
206-565-6600
Fax: 206-565-7265
www.APAWOOD.org

SPECIFICATIONS DATA SOURCES -- TECHNICAL DATA continued

06000 WOOD continued

__ Specifications, technical standards for architectural woodwork.

> AWI -- Architectural Woodwork Institute
> 1952 Isaac Newton Square
> Reston, VA 20190
> 703-733-0600
> Fax: 703-733-0584
> www.AWINET.org

__ AWPA Book of Standards.

> AWPA -- American Wood-Preservers' Association
> 3246 Fall Creek Highway, Suite 190
> Granbury, TX 76049
> 817-326-6300
> Fax: 817-326-6306

__ Standard Specifications for redwood lumber.

> California Redwood Lumber
> CRA -- California Redwood Association
> 405 Enfrente Drive, Suite 200
> Novato, CA 94949
> 415-382-0662
> Fax: 415-382-8531
> E-mail: grovercf@slip.net

__ Hardwood plywood design standards.

> HPVA -- Hardwood Plywood & Veneer Association
> 1825 Michael Faraday Drive
> Reston, VA 20190-5350
> 703-435-2900
> Fax: 703-435-2537
> E-mail: hpval@erols.com
> www.erols.com/hpva

__ National Design Specification for Wood Construction.

> National Forest Products Ass'n.
> 1250 Connecticut Ave. N.W.
> Washington, DC 20036

__ Standard Grading Rules for West Coast Lumber.

> WCLIB -- West Coast Lumber Inspection Bureau
> P.O. Box 23145
> Portland, OR 97281
> 503-639-0651

SPECIFICATIONS DATA SOURCES --
TECHNICAL DATA continued

A-8

07000 THERMAL & MOISTURE PROTECTION

__ Design standards, technical literature, and specifications for curtain walls; store fronts and entrances; glazing and skylights.

AAMA -- American Architectural Manufacturers Association
1827 Walden Office Square, Suite 104
Schaumburg, IL 60173
847-303-5664
Fax: 847-303-5774
E-mail: webmaster@aamanet.org
www.AAMANET.org

ARMA -- Asphalt Roofing Manufacturers Association
6000 Executive Boulevard, Suite 201
Rockville, MD 20852-3803
301-231-9050
Fax 301-881-6572

CSSB -- Cedar Shake and Shingle Bureau
515 116th Ave. N.E., Suite 275
Bellevue, WA 98004
206-453-1323
Fax: 206-455-1314
E-mail: cssblind@nwlink.com
www.CEDARBUREAU.org

NRCA -- National Roofing Contractors Association
Technical Services
10255 W. Higgins Road, Suite 600
Rosemont, IL 60018
847-299-9070
1-800-323-9545
Fax: 847-299-1183
E-mail: jenhofme@roofonline.org
www.ROOFONLINE.org

SWRI -- Sealant, Waterproofing & Restoration Institute
3101 Broadway, Suite 585
Kansas City, MO 64111
816-561-8230
Fax: 816-561-7765

SPRI -- Single Ply Roofing Institute
175 Highland Ave.
Needham, MA 02194
617-444-0242
Fax: 617-444-6111
E-mail: spri@spri.org
www.SPRI.org

SPECIFICATIONS DATA SOURCES -- TECHNICAL DATA continued

08000 GLAZING

GANA -- Glass Association of North America
3310 S.W. Harrison Street
Topeka, KS 66611-2279
913-266-7013
Fax: 913-266-0272

08700 HARDWARE

BHM -- Builders Hardware Manufacturers Association
355 Lexington Ave., 17th Floor
New York, NY 10017
212-661-4261
Fax: 212-370-9047

DHI -- Door and Hardware Institute
14170 Newbrook Drive
Chantilly, VA 20151-2232
703-222-2010
Fax: 703-222-2410
www.dhi.org

09000 FINISHES

CRI -- Carpet and Rug Institute
310 Holiday Drive, P.O. Box 2048
Dalton, GA 30722-2048
706-278-3176
1-800-882-8846
Fax 706-278-8835

CPI -- Chicago Plastering Institute
6547 North Avondale Ave., #202
Chicago, Il 60631
773-774-4500

_ Design standards and technical information for ceilings and acoustical treatments

COSCA -- Ceilings & Interior Systems Construction Association
1500 Lincoln Highway, Suite 202
St. Charles, IL 60174
630-584-1919
Fax: 630-584-2003
E-mail: 75031.2577@compuserve.com

SPECIFICATIONS DATA SOURCES -- TECHNICAL DATA continued

A-10

09000 FINISHES continued

CTIOA -- Ceramic Tile Institute of America
12061 W. Jefferson
Culver City, CA 90230-6219
310-574-7800
Fax: 310-821-4655

GA -- Gypsum Association
810 First Street N.E., Suite 510
Washington, DC 20002
202-289-5440
Fax: 202-289-3707

IILP -- International Institute for Lath and Plaster
820 Transfer Road, Suite 34
St. Paul, MN 55114
612-645-0208
Fax: 612-645-0209

TCA -- Tile Council of America, Inc.
P.O. Box 1787
Clemson, SC 29633
864-646-8453
Fax: 864-646-2821
Email: tcalink@carol.net
www.TILEUSA.com/

NTMA -- National Terrazzo and Mosaic Association
3166 Des Plaines Avenue, Suite 121
Des Plaines, IL 60018
847-635-7744
Fax: 847-635-9127
1-800-323-9736

N.W.FA -- National Wood Flooring Association
233 Old Maremac Station Road
Manchester, MO 63021-5310
314-391-5161
Fax: 314-391-6137
Toll free phone: 800-422-4556
e-mail: nwfa@aol.com
www.woodfloors.org/

SPECIFICATIONS DATA SOURCES -- TECHNICAL DATA continued

ALTERNATIVE CONSTRUCTION AND ENERGY SYSTEMS

ASES -- American Solar Energy Society
2400 Central Avenue, Suite G-1
Boulder, CO 80301
303-443-3130
Fax: 303-443-3212
E-mail: ASES@ASES.org
www.ASES.org/solar

AUA -- American Underground Construction Association
511 11th Avenue South, Suite 248
Minneapolis, MN 55415
612-339-5403
Fax: 612-339-3207

AWEA -- American Wind Energy Association
122 C St. N.W., 4th Floor
Washington, DC 20001
202-383-2500
Fax: 202-383-2505
Fax on Demand: 800-634-4299

BETEC -- Building Environmental and Thermal Envelope Council
National Institute of Building Sciences
1201 L Street N.W., Suite 400
Washington, DC 20005-4024
202-289-7800
Fax: 202-289-1092

CRBT -- Center for Resourceful Building Technology
P.O. Box 100
Missoula, MT 59806
406-549-7678
Fax: 406-549-4100

CREST -- Center for Renewable Energy and Sustainable Technology
1200 18th Street N.W., #900
Washington, DC 20036
202-530-2202
Fax: 202-887-0497

EEBA -- Energy Efficient Buildings Association
2950 Metro Drive, Suite 108
Minneapolis, MN 55425-1560
612-851-9940
Fax: 612-851-9507

PSIC -- Passive Solar Industries Council
1511 K Street N.W., Suite 600
Washington, DC 20005
202-628-7400
Fax: 202-393-5043
E-mail: psicouncil@aol.com
www.psic.org

SPECIFICATIONS DATA SOURCES --
FAILURE PREVENTION READING LIST   A-12

__ Construction Disasters -- Design Failures, Causes & Prevention, by Steven S. Ross. McGraw-Hill Book Co. ISBN 0-07-053865-4.

__ Construction Failures, Cushman, Richter, Rivelis, Editors. John Wiley & Sons. (Brand new edition.)

__ Structural & Foundation Failures, by Barry B. LePatner & Sidney M. Johnson. McGraw-Hill Book Co. ISBN 0-07-032584-7.

__ Construction Sealants and Adhesives, by Julian R. Panek and John P. Cook. John Wiley & Sons. ISBN 0-471-09360-2.

__ The Professional Practice of Architectural Detailing, by Wakita & Linde. John Wiley & Sons. ISBN 0-471-91715-X.

__ Roofing Concepts & Principles -- A Practical Approach to Roofing. Paul Tente, Paul Tente Associates, Box 6819, Colorado Springs, CO 80934.

__ NRCA Roofing & Waterproofing Manual (Includes NRCA recommended mended construction details), National Roofing Contractors Association, 8600 Bryn Mawr Ave., Chicago, IL 60631.

__ Avoiding Liability in Architecture, Design & Construction, edited by Robert F. Cushman. John Wiley & Sons, Inc. Wiley-Interscience. ISBN 0-471-09579-6.

For construction design, drafting, and spec aids we recommend:

__ Directory of Publications of Loss Prevention Aids. (Free list of books, manuals, video and audio programs. These are helpful to all design professionals, not just the ASFE members.)

   Association of Soil and Foundations Engineers
   8811 Colesville Rd. Suite 225
   Silver Spring, MD 20910

__ The CSI Catalog of Services and Publications. (Free catalog of manuals, self-study aids, cassette tapes, and model specifications sections and technical aids. A priceless source of data.)

   Construction Specifications Institute
   601 Madison St.
   Alexandria, VA 22314

__ BRP Publications "Reports for Sale."

   National Academy Press, The Building Research Board
   2101 Constitution Ave., N.W.
   Washington, DC 20418

__ National Research Council of Canada "List of Publications." (Lots of extremely useful research data on construction of all kinds.)

   Publications Section, Division of Building Research
   National Research Council of Canada
   Ottawa, Ontario, K1A 0R6

Notes:

Notes:

Notes:

Notes:

Notes:

Notes:

Notes:

Notes:

Notes:

Notes:

Notes:

Notes:

Notes: